The Beginner's Guide to Engineering: Chemical Engineering

quantum scientific publishing

The Beginner's Guide to Engineering: Chemical Engineering

John T. Stimus

quantum scientific publishing

The Beginner's Guide to Engineering: Chemical Engineering

ISBN-13: 978-1492965046
ISBN-10: 1492965049

Published by quantum scientific publishing

Pittsburgh, PA

Cover design by Scott Sheariss

Table of Contents

Unit One

Unit Two

Unit Three

Appendix

Unit One

Section 1.1 – Engineering vs. Science

Section Objective

- Explain the difference between engineering and science

Science and Engineering – what is the difference?

Science and engineering are different but related disciplines. **Science** involves developing a body of knowledge about a topic by doing experiments. Scientists do experiments using the scientific method, which is a disciplined, systematic approach to developing and answering questions, to add to the body of knowledge. In order for a scientific finding to be accepted as true, the experiment must be repeated by other scientists with the same result.

Dictionary.com defines science the following ways:

> a branch of knowledge or study dealing with a body of facts or truths systematically arranged and showing the operation of general laws: *the mathematical sciences.*
>
> systematic knowledge of the physical or material world gained through observation and experimentation.
>
> any of the branches of natural or physical science.
>
> systematized knowledge in general.
>
> knowledge, as of facts or principles; knowledge gained by systematic study.
>
> a particular branch of knowledge.
>
> skill, esp. reflecting a precise application of facts or principles; proficiency.

Technicians freezing samples in liquid nitrogen

Image courtesy of United States Department of Agriculture

Engineering is the application of scientific ideas to real life. Engineers use scientific principles to develop the tools and products used by people just like you. Engineers design buildings, cars, eyeglasses, furniture, electronics, cleaning products, medications, medical equipment and many other things.

Dictionary.com defines engineering the following ways:

> the art or science of making practical application of the knowledge of pure sciences, as physics or chemistry, as in the construction of engines, bridges, buildings, mines, ships, and chemical plants.
>
> the action, work, or profession of an engineer.
>
> skillful or artful contrivance; maneuvering.

As we can see by these definitions, science and engineering are closely related. One is neither better nor worse than the other, they are just different. Scientists develop the knowledge that engineers apply to solve real life problems.

Science	Engineering
Asks why?	Asks how?
Fundamental	Practical
Investigate phenomena	Create solutions

The Relationship between Science and Engineering

Engineering and science have existed since the dawn of history. Early scientists were simply people who tried to explain how nature works. They would conduct very simple experiments and make predictions to test their ideas. Many of the major principles in science date back to the ancient Greek philosophers, who gave us such concepts as atoms, elements, matter, mass, volume, and density.

Early engineers were what we would consider today to be tinkerers or inventors. They were people who built items to make their lives easier by using their knowledge of science. These early engineers often built by trial and error because they lacked formal training in engineering and design concepts. Through persistence and a desire to make use of their knowledge of nature, these early engineers helped to advance mankind through the creation of technology.

An example of the relationship between science and engineering can be seen in the history of fire. Primitive scientists studied natural fires caused by lightning or volcanic activity and learned how fires started and spread. Using this knowledge, primitive engineers learned how to build and maintain fires. These fires could then be used for light, warmth, cooking, and even the smelting of metals. As scientists learned about different materials that are combustible, like oil, coal and wax, engineers could find new ways to make use of these materials for small or large fires.

Chemistry vs. Chemical Engineering

Chemistry is the branch of science that studies the structure and properties of matter. Chemistry studies how matter is arranged and how it can be changed by chemical reactions. Chemists study what matter is made of and how it can be changed physically and chemically into new substances.

Chemical engineering is the field of engineering that applies the knowledge of chemistry to make new products. Chemical engineers perform many different jobs, including research, design, and management. Some engineers perform research about new products and processes to make those products. Other engineers design chemical plants to product chemicals. Finally, some engineers are involved in the management of chemical plants.

As chemistry develops as a science, engineers are provided with new information to use in the production of current products as well as information that can be used to make new products. The technology and processes used by chemical engineers may lead to further investigations by chemists, which in turn eventually lead to new knowledge about chemistry. In this way, chemistry and chemical engineering work together to further our understanding and use of matter in the world around us.

Concept Reinforcement

1. What kind of work do scientists do?
2. What kind of work do engineers do?
3. Are the following items science or engineering?
    a. Discovering a new chemical
    b. Designing a plant to produce a chemical.
    c. Building a filter to clean water.
    d. Discovering a new drug.
    e. Finding the melting point of a plastic.
    f. Finding new ways to use a plastic.

Section 1.2 – Systems of Units and Unit Conversions

Section Objective

- List and describe systems of units and unit conversions

Units

In engineering, there must be a consistent way to quantify the physical world around us. Engineers use units as the consistent basis of their measurement systems. A **unit** is a set amount of a measurement. Feet, inches, and meters are all units of measuring length. Engineers must know which type of unit is being used to complete a project. If one engineer is using a different set of units from another, they may end up designing parts that are not compatible with one another. For example, if a chemical engineer is measuring the length of a pipe in meters while another one is using feet, they will have to be very careful to specify which unit they are using to prevent confusion.

In ancient times, people used systems of measurement that were not always easily reproducible. They would use parts of their bodies as a way to measure things. For instance, length was measured in cubits. A cubit was the distance from the tip of the finger to the elbow. If you look at the people you know, you will notice that the lengths of their arms vary quite a bit. Another measurement that had a lot of variability was an acre. We now define an acre as 4047 square yards. In medieval times, an acre was the amount of land that one man with one ox could plow in a day. As you can see, this can vary based on the man, the ox, and the type of land. So how do you standardize measurements? A way to standardize the way things were measured had to be devised.

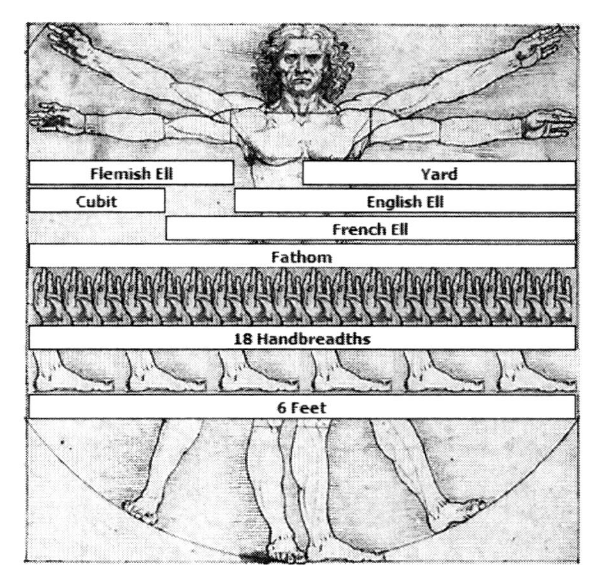

An adaptation of DaVinci's Vitruvian Man showing various measures of length

The Metric System

In the late 1700s, the French started to standardize measurements with the development of the **metric system**. To determine length, they divided the earth into quadrants, the distance from the equator to the poles. They then accurately measured the quadrant from the equator to the North Pole passing through Paris. This distance was divided into ten million equal parts. Each equal part was defined as a **meter**. This was called the meridional definition. The meter was then used to define other quantities. Volume was standardized to the **liter**. This was equivalent to one thousandth of a cubic meter. Mass was standardized to the **gram**. A gram is one thousandth of the mass of a liter of water. Measurements like these can now be reproduced by any scientist or engineer anywhere in the world. So if you build a machine in one city, it can be easily reproduced in any other city in the world.

Further standardization

In 1889 a "prototype" meter bar was developed. This bar was made of iridium and platinum, and was used as the primary comparison for determining the meter. In 1960, a meter was defined as 1,650,763.73 wavelengths of the orange-red radiation of krypton 86 under specified conditions. In 1983 the meter was finally defined as the length of the path travelled by light in a vacuum during a time interval of 1/299,792,458 of a second.

The metric system standardized the way we measure things. This, along with the fact the metric system is based on factors of ten, makes use of the metric system very easy. Currently there are seven base units for the measurement of most quantities. These measurements are:

Table 1: Metric Units

Name	Symbol	Quantity
Meter	m	Length
Kilogram	kg	Mass
Second	S	Time
Ampere	A	Electric Current
Kelvin	K	Temperature
Candela	cd	Luminous intensity
Mole	mol	Amount of a substance

Converting between Metric Units

A **prefix** may be added to any unit to produce some multiple or division of that unit. These multiples and divisions are all power of ten. To convert a measurement from the base unit (grams, meters, liters, etc.) to a multiple (deca, hecto, kilo, etc.), we need to divide by the factor of the multiple (10, 100, 1,000, etc.). We divide when converting to a multiple of the unit because the base units are smaller.

Table 2: Prefixes for Multiples of Metric Units

Name	Unit (meter, liter, gram, etc.)	Deca	Hecto	Kilo	Mega	Giga
Symbol		da	h	k	M	G
Factor	1	10	100	1,000	1,000,000	1,000,000,000

Example 1: How many kilograms are there in 563 grams?

Using Table 2, we see that a kilogram is 1,000 grams. This tells us that we need to divide the number of units by 1,000 to find the number of kilograms.

563 g / 1,000 = 0.050 kg

Answer: 0.563 kg

Notice that converting from a base unit to a multiple requires division. Also, notice that it takes fewer kilograms (0.563 kg) to equal the same amount in grams (563 g). This is because kilograms are larger, and it takes fewer kilograms to equal the same mass in grams. If we converted from kilograms back to grams, we would multiply the number of kilograms by 1,000.

0.563 kg • 1,000 = 563 g

To convert a measurement from the base unit (grams, meters, liters, etc.) to a division (deci, centi, milli, etc), we need to multiply by the factor of the division (10, 100, 1,000, etc.). We multiply when converting to a division of the unit because the base units are larger than the divisions. If we convert divisions back to base units, we divide the number of divisions by its factor to find the number of units.

Table 3: Prefixes for Divisions of Metric Units

Name	Unit (meter, liter, gram, etc.)	Deci	Centi	Milli	Micro	Nano
Symbol		d	c	m	μ	n
Factor	1	0.1	0.01	0.001	0.0000001	0.0000000001

Example 2: How many meters are there in 750 centimeters?

Using Table 3, we see that a centimeter is 0.01 meters. This tells us that there are 100 centimeters in a meter. To convert from centimeters (the division) to meters (the base unit), we divide the number of centimeters by 100.

750 cm / 100 = 7.50 m

Answer: 7.50 m

Notice that there are fewer meters (0.750 m) than centimeters (750 cm). This is because centimeters are smaller than meters, and it would take many centimeters to measure a distance that can be measured in just a few meters. If we convert the base unit back to centimeters, we need to multiply by a factor of 100.

7.50 m • 100 = 750 cm

Converting from US Standard to Metric Units

Prior to this standardization, there were and still are many different conversion factors for moving from one type of units to another type of units. For countries such as the US, where the metric system is not widely used, conversion factors such as the following are used:

Table 4: Conversion Factors from US Units to Metric Units:

Quantity	US Unit	Metric Unit	Multiply by
Length	Foot	Meter	0.3048
Mass	Pound	Kilogram	0.45
Volume	Gallon	Liter	3.785
Length	Yards	Meters	0.9144
Speed	Miles per hour	Kilometers per hour	1.609

To convert a US Standard unit to a metric unit, simply multiply the number of units by the conversion factor listed in Table 4. To convert a metric unit to US Standard, we simply divide by the same factor.

Example 3: How many kilograms are 5 pounds equal to?

Using Table 4, we see that 1 pound is equal to 0.45 Kilograms. To convert 5 pounds to kilograms, we simple multiply the number of pounds by 0.45.

5 pounds • 0.45 = 2.25 kg

Answer: 2.25 kg

To convert this measurement back to pounds, we divide 2.25 kg by 0.45.

2.25 kg / 0.45 = 5 pounds

Concept Reinforcement

1. What is a unit?
2. How many kilometers are there in 3,160 meters?
3. How many liters are there in 50 milliliters?
4. How many liters are 12 gallons equal to?
5. How many meters are 10 feet equal to?

Section 1.3 – Structure of an Atom

Section Objective

- Describe the structure of the atom

What is an atom?

An **atom** is the building block of all matter. An atom is composed of a small central **nucleus**, surrounded by a cloud of negatively charged particles called **electrons**. The nucleus itself is composed of two particles: a positively charged **proton**, and a neutrally charged **neutron**.

Image of a nucleus

History of Atomic Structure

The Greeks had some of the oldest theories of the atom. In Greek atom means "uncuttable." If you took a saw and cut piece of silver in half, and kept halving the next piece until it was too small to cut anymore, that would be an atom.

Throughout the next millennia different people added to the understanding of the atom and its structure. In 1808 John Dalton proposes his Atomic Theory. His theory states:

> *Matter is made of tiny particles called atoms.*
>
> *All atoms of a given element are identical*
>
> *Atoms of different elements have different properties*
>
> *Atoms are neither created nor destroyed in chemical reactions.*
>
> *Atoms of different elements form compounds in whole number ratios*

While working with cathode ray tubes in 1897, J.J. Thompson discovered the electron, and proved that atoms can be divided further. In 1909, Ernest Rutherford fired helium nuclei at a piece of incredibly thin gold foil and discovered that the atom is mostly made up of empty space. His helium nuclei passed through the gold shield easily, only being deflected a small percentage of the time. He hypothesized that the electrons move around the central nucleus the way planets revolve around a star.

In 1913, Niels Bohr refined Rutherford's Theory, showing that electrons were confined to quantized **orbitals**. The electrons could move between orbitals if it absorbed or released energy, but could not freely move between these orbits. In 1926, Erwin Schroedinger showed that electrons have properties of both a particle and a wave.

With this work, and the continued work of many other scientists, we have a modern theory for the structure of the atom. We now know that atoms are "cuttable." They are made up of several different types of subatomic particles. Protons and neutrons are made up of elementary particles called **quarks**. Electrons are already elementary particles and form an electron "cloud" around the nucleus being held there by the electromagnetic force. This electron cloud forms a three dimensional wave with respect to the nucleus. This wave can be represented mathematically as a function of the probability where the electron will be.

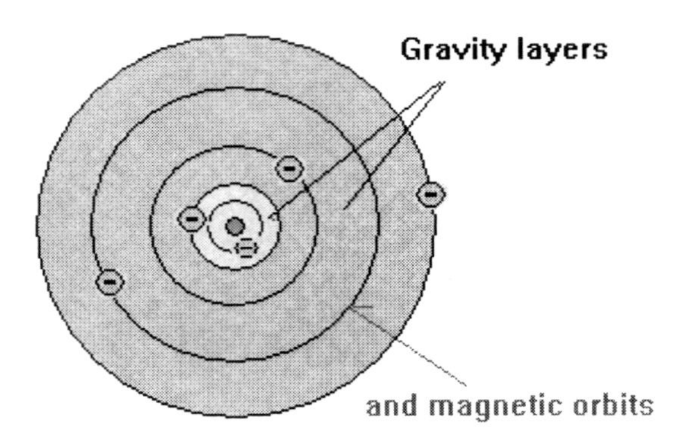

Diagram of orbitals

Concept Reinforcement

1. What is an atom?
2. Who developed the first theories about the atom?
3. Who showed that the atom was mostly empty space?
4. What elementary particles make up a proton?
5. What is the charge of:
   a. A proton?
   b. A neutron?
   c. An electron?

Section 1.4 – Elements and the Periodic Table

Section Objective

- Explain the elements and the periodic table of the elements

Elements

What are we made of? What is the basic structure of matter? In ancient Greece, Plato proposed that there were four basic elements, air, fire, earth and water. These elements all had polyhedral shapes. Fire was tetrahedral (4-sided), earth was cubic (6-sided), air was octahedral (8-sided), water was icosahedral (20-sided). Aristotle later added quintessence, the substance that makes up the heavens. His definition of the element was "one of those bodies into which other bodies can be decomposed and which itself is not capable of being divided into other." Over the next almost 2000 years, scientist questioned this, but due to lack of experimental evidence, could not arrive at a proper explanation.

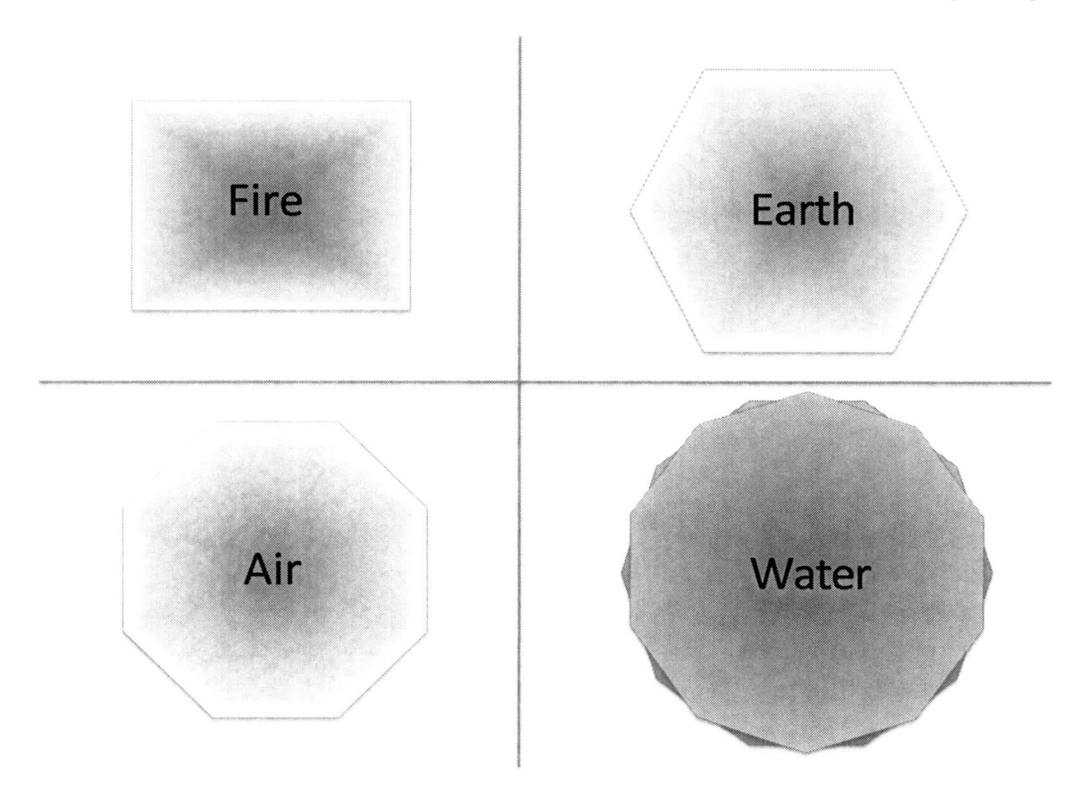

In 1789, Antoine Lavoisier published the Elements of Chemistry which had 33 elements. He included caloric and light as elements in his book. By 1869 Dmitri Mendeleev had 66 elements in his periodic table. To organize his elements, Mendeleev placed them in a table, based on their **atomic masses**. This organization worked well until Henry Mosley placed them in order by **atomic number** (number of protons in the nucleus) as opposed to mass.

The Modern Periodic Table

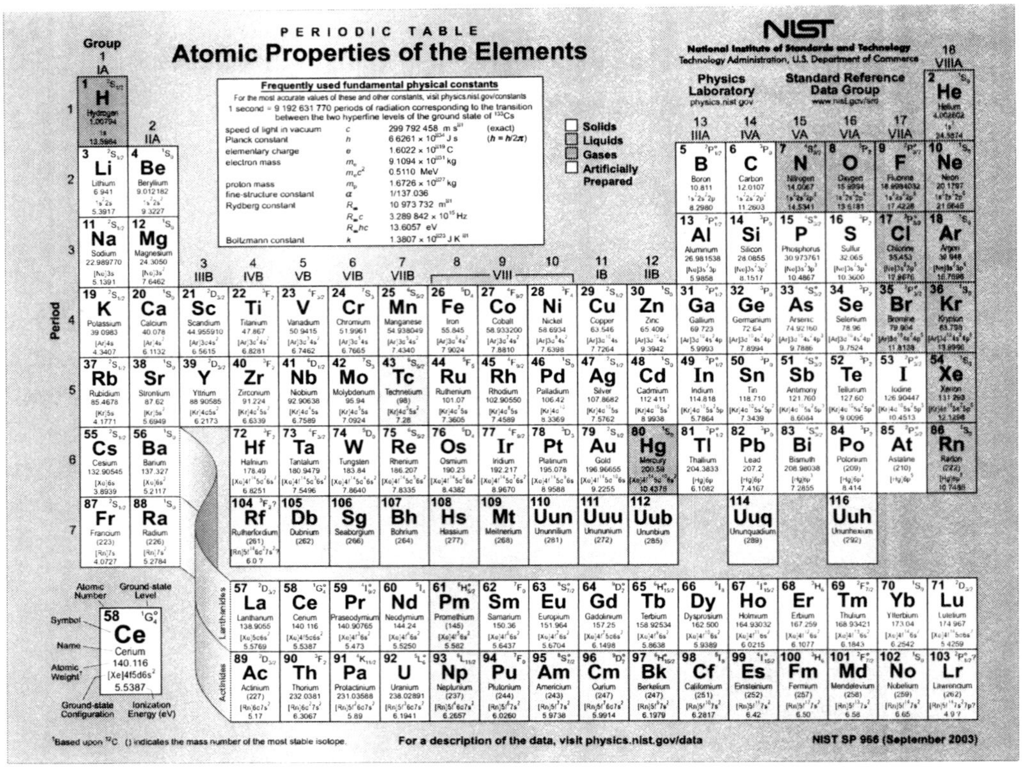

PERIODIC TABLE

Atomic Properties of the Elements

NIST — National Institute of Standards and Technology, Technology Administration, U.S. Department of Commerce

Physics Laboratory, physics.nist.gov — Standard Reference Data Group, www.nist.gov/srd

Frequently used fundamental physical constants

For the most accurate values of these and other constants, visit physics.nist.gov/constants

1 second = 9 192 631 770 periods of radiation corresponding to the transition between the two hyperfine levels of the ground state of ^{133}Cs

Constant	Symbol	Value	
speed of light in vacuum	c	299 792 458 m s^{-1}	(exact)
Planck constant	h	6.6261 × 10^{-34} J s	($\hbar = h/2\pi$)
elementary charge	e	1.6022 × 10^{-19} C	
electron mass	m_e	9.1094 × 10^{-31} kg	
	$m_e c^2$	0.5110 MeV	
proton mass	m_p	1.6726 × 10^{-27} kg	
fine-structure constant	α	1/137.036	
Rydberg constant	R_∞	10 973 732 m^{-1}	
	$R_\infty c$	3.289 842 × 10^{15} Hz	
	$R_\infty hc$	13.6057 eV	
Boltzmann constant	k	1.3807 × 10^{-23} J K^{-1}	

Solids / Liquids / Gases / Artificially Prepared

Groups: 1 IA, 2 IIA, 3 IIIB, 4 IVB, 5 VB, 6 VIB, 7 VIIB, 8–10 VIII, 11 IB, 12 IIB, 13 IIIA, 14 IVA, 15 VA, 16 VIA, 17 VIIA, 18 VIIIA

Key: Atomic Number, Ground-state Level, Symbol, Name, Atomic Weight†, Ground-state Configuration, Ionization Energy (eV) — e.g. 58 $^1G^\circ_4$ Ce Cerium 140.116 $[Xe]4f5d6s^2$ 5.5387

No.	Symbol	Name	Atomic Weight	Ground-state Configuration	Ionization Energy (eV)
1	H	Hydrogen	1.00794	$1s$	13.5984
2	He	Helium	4.002602	$1s^2$	24.5874
3	Li	Lithium	6.941	$1s^2 2s$	5.3917
4	Be	Beryllium	9.012182	$1s^2 2s^2$	9.3227
5	B	Boron	10.811	$1s^2 2s^2 2p$	8.2980
6	C	Carbon	12.0107	$1s^2 2s^2 2p^2$	11.2603
7	N	Nitrogen	14.0067	$1s^2 2s^2 2p^3$	14.5341
8	O	Oxygen	15.9994	$1s^2 2s^2 2p^4$	13.6181
9	F	Fluorine	18.9984032	$1s^2 2s^2 2p^5$	17.4228
10	Ne	Neon	20.1797	$1s^2 2s^2 2p^6$	21.5646
11	Na	Sodium	22.989770	$[Ne]3s$	5.1391
12	Mg	Magnesium	24.3050	$[Ne]3s^2$	7.6462
13	Al	Aluminum	26.981538	$[Ne]3s^2 3p$	5.9858
14	Si	Silicon	28.0855	$[Ne]3s^2 3p^2$	8.1517
15	P	Phosphorus	30.973761	$[Ne]3s^2 3p^3$	10.4867
16	S	Sulfur	32.065	$[Ne]3s^2 3p^4$	10.3600
17	Cl	Chlorine	35.453	$[Ne]3s^2 3p^5$	12.9676
18	Ar	Argon	39.948	$[Ne]3s^2 3p^6$	15.7596
19	K	Potassium	39.0983	$[Ar]4s$	4.3407
20	Ca	Calcium	40.078	$[Ar]4s^2$	6.1132
21	Sc	Scandium	44.955910	$[Ar]3d4s^2$	6.5615
22	Ti	Titanium	47.867	$[Ar]3d^2 4s^2$	6.8281
23	V	Vanadium	50.9415	$[Ar]3d^3 4s^2$	6.7462
24	Cr	Chromium	51.9961	$[Ar]3d^5 4s$	6.7665
25	Mn	Manganese	54.938049	$[Ar]3d^5 4s^2$	7.4340
26	Fe	Iron	55.845	$[Ar]3d^6 4s^2$	7.9024
27	Co	Cobalt	58.933200	$[Ar]3d^7 4s^2$	7.8810
28	Ni	Nickel	58.6934	$[Ar]3d^8 4s^2$	7.6398
29	Cu	Copper	63.546	$[Ar]3d^{10} 4s$	7.7264
30	Zn	Zinc	65.409	$[Ar]3d^{10} 4s^2$	9.3942
31	Ga	Gallium	69.723	$[Ar]3d^{10} 4s^2 4p$	5.9993
32	Ge	Germanium	72.64	$[Ar]3d^{10} 4s^2 4p^2$	7.8994
33	As	Arsenic	74.92160	$[Ar]3d^{10} 4s^2 4p^3$	9.7886
34	Se	Selenium	78.96	$[Ar]3d^{10} 4s^2 4p^4$	9.7524
35	Br	Bromine	79.904	$[Ar]3d^{10} 4s^2 4p^5$	11.8138
36	Kr	Krypton	83.798	$[Ar]3d^{10} 4s^2 4p^6$	13.9996
37	Rb	Rubidium	85.4678	$[Kr]5s$	4.1771
38	Sr	Strontium	87.62	$[Kr]5s^2$	5.6949
39	Y	Yttrium	88.90585	$[Kr]4d5s^2$	6.2173
40	Zr	Zirconium	91.224	$[Kr]4d^2 5s^2$	6.6339
41	Nb	Niobium	92.90638	$[Kr]4d^4 5s$	6.7589
42	Mo	Molybdenum	95.94	$[Kr]4d^5 5s$	7.0924
43	Tc	Technetium	(98)	$[Kr]4d^5 5s^2$	7.28
44	Ru	Ruthenium	101.07	$[Kr]4d^7 5s$	7.3605
45	Rh	Rhodium	102.90550	$[Kr]4d^8 5s$	7.4589
46	Pd	Palladium	106.42	$[Kr]4d^{10}$	8.3369
47	Ag	Silver	107.8682	$[Kr]4d^{10} 5s$	7.5762
48	Cd	Cadmium	112.411	$[Kr]4d^{10} 5s^2$	8.9938
49	In	Indium	114.818	$[Kr]4d^{10} 5s^2 5p$	5.7864
50	Sn	Tin	118.710	$[Kr]4d^{10} 5s^2 5p^2$	7.3439
51	Sb	Antimony	121.760	$[Kr]4d^{10} 5s^2 5p^3$	8.6084
52	Te	Tellurium	127.60	$[Kr]4d^{10} 5s^2 5p^4$	9.0096
53	I	Iodine	126.90447	$[Kr]4d^{10} 5s^2 5p^5$	10.4513
54	Xe	Xenon	131.293	$[Kr]4d^{10} 5s^2 5p^6$	12.1298
55	Cs	Cesium	132.90545	$[Xe]6s$	3.8939
56	Ba	Barium	137.327	$[Xe]6s^2$	5.2117
57	La	Lanthanum	138.9055	$[Xe]5d6s^2$	5.5769
58	Ce	Cerium	140.116	$[Xe]4f5d6s^2$	5.5387
59	Pr	Praseodymium	140.90765	$[Xe]4f^3 6s^2$	5.473
60	Nd	Neodymium	144.24	$[Xe]4f^4 6s^2$	5.5250
61	Pm	Promethium	(145)	$[Xe]4f^5 6s^2$	5.582
62	Sm	Samarium	150.36	$[Xe]4f^6 6s^2$	5.6437
63	Eu	Europium	151.964	$[Xe]4f^7 6s^2$	5.6704
64	Gd	Gadolinium	157.25	$[Xe]4f^7 5d6s^2$	6.1498
65	Tb	Terbium	158.92534	$[Xe]4f^9 6s^2$	5.8638
66	Dy	Dysprosium	162.500	$[Xe]4f^{10} 6s^2$	5.9389
67	Ho	Holmium	164.93032	$[Xe]4f^{11} 6s^2$	6.0215
68	Er	Erbium	167.259	$[Xe]4f^{12} 6s^2$	6.1077
69	Tm	Thulium	168.93421	$[Xe]4f^{13} 6s^2$	6.1843
70	Yb	Ytterbium	173.04	$[Xe]4f^{14} 6s^2$	6.2542
71	Lu	Lutetium	174.967	$[Xe]4f^{14} 5d6s^2$	5.4259
72	Hf	Hafnium	178.49	$[Xe]4f^{14} 5d^2 6s^2$	6.8251
73	Ta	Tantalum	180.9479	$[Xe]4f^{14} 5d^3 6s^2$	7.5496
74	W	Tungsten	183.84	$[Xe]4f^{14} 5d^4 6s^2$	7.8640
75	Re	Rhenium	186.207	$[Xe]4f^{14} 5d^5 6s^2$	7.8335
76	Os	Osmium	190.23	$[Xe]4f^{14} 5d^6 6s^2$	8.4382
77	Ir	Iridium	192.217	$[Xe]4f^{14} 5d^7 6s^2$	8.9670
78	Pt	Platinum	195.078	$[Xe]4f^{14} 5d^9 6s$	8.9588
79	Au	Gold	196.96655	$[Xe]4f^{14} 5d^{10} 6s$	9.2255
80	Hg	Mercury	200.59	$[Xe]4f^{14} 5d^{10} 6s^2$	10.4375
81	Tl	Thallium	204.3833	$[Hg]6p$	6.1082
82	Pb	Lead	207.2	$[Hg]6p^2$	7.4167
83	Bi	Bismuth	208.98038	$[Hg]6p^3$	7.2855
84	Po	Polonium	(209)	$[Hg]6p^4$	8.414
85	At	Astatine	(210)	$[Hg]6p^5$	
86	Rn	Radon	(222)	$[Hg]6p^6$	10.7485
87	Fr	Francium	(223)	$[Rn]7s$	4.0727
88	Ra	Radium	(226)	$[Rn]7s^2$	5.2784
89	Ac	Actinium	(227)	$[Rn]6d7s^2$	5.17
90	Th	Thorium	232.0381	$[Rn]6d^2 7s^2$	6.3067
91	Pa	Protactinium	231.03588	$[Rn]5f^2 6d7s^2$	5.89
92	U	Uranium	238.02891	$[Rn]5f^3 6d7s^2$	6.1941
93	Np	Neptunium	(237)	$[Rn]5f^4 6d7s^2$	6.2657
94	Pu	Plutonium	(244)	$[Rn]5f^6 7s^2$	6.0260
95	Am	Americium	(243)	$[Rn]5f^7 7s^2$	5.9738
96	Cm	Curium	(247)	$[Rn]5f^7 6d7s^2$	5.9914
97	Bk	Berkelium	(247)	$[Rn]5f^9 7s^2$	6.1979
98	Cf	Californium	(251)	$[Rn]5f^{10} 7s^2$	6.2817
99	Es	Einsteinium	(252)	$[Rn]5f^{11} 7s^2$	6.42
100	Fm	Fermium	(257)	$[Rn]5f^{12} 7s^2$	6.50
101	Md	Mendelevium	(258)	$[Rn]5f^{13} 7s^2$	6.58
102	No	Nobelium	(259)	$[Rn]5f^{14} 7s^2$	6.65
103	Lr	Lawrencium	(262)	$[Rn]5f^{14} 7s^2 7p?$	4.9 ?
104	Rf	Rutherfordium	(261)	$[Rn]5f^{14} 6d^2 7s^2 ?$	6.0 ?
105	Db	Dubnium	(262)		
106	Sg	Seaborgium	(266)		
107	Bh	Bohrium	(264)		
108	Hs	Hassium	(277)		
109	Mt	Meitnerium	(268)		
110	Uun	Ununnilium	(281)		
111	Uuu	Unununium	(272)		
112	Uub	Ununbium	(285)		
114	Uuq	Ununquadium	(289)		
116	Uuh	Ununhexium	(292)		

Lanthanides (57–71); Actinides (89–103)

†Based upon ^{12}C. ( ) indicates the mass number of the most stable isotope.

For a description of the data, visit physics.nist.gov/data

NIST SP 966 (September 2003)

The modern **periodic table** is an organized method of presenting the elements. In the modern definition, an **element** is a pure chemical substance composed of atoms with the same number of protons. Elements in the periodic table are listed in order of their atomic number, which is the number of protons found in the nucleus of each element.

The vertical columns of the periodic table are called **groups**. These groups of elements are grouped by similar physical and chemical properties. The horizontal rows of the periodic table are called **periods**. The one thing that periods have in common is that energies of the outermost electrons are similar. Elements on the left side of each period have very few electrons in the outer layer of their electron shells, while elements on the right side of the table have full or almost full electron shells. The number of electrons in the outer shell of an element is the main factor in determining the chemical properties of the element.

Periodic Families

All of the groups in the periodic table can also be arranged into **families**. Most families are made up of just one group of elements, but several families contain more than one group. Each family of elements share common properties and atomic structures. Below are listed the families of the periodic table:

Alkali Metals

Group 1

Very reactive

One electron in outer shell

Oxidation number +1

Ductile

Malleable

Good conductors of heat and electricity

Explode if exposed to water

Alkali Earth Metals

Group 2

Very reactive

React violently with water

Two electrons in outer shell

Oxidation number +2

Transition Metals

Group 3-12

Ductile

Malleable

Conducts heat and electricity

Valence electrons in more than one shell

Several oxidation states

Other Metals

Groups 13, 14, 15

Ductile

Malleable

Valence electrons only ion outer shell

Non-variable oxidation states

Metalloids

Parts of groups 13-16 – form a "staircase" from the top of group 13 to the bottom of group 16

Include Boron, Silicon, Germanium, Arsenic, Antimony, Tellurium, and Polonium

Have properties of both metals and non-metals

Are often used as semiconductors in electronics

Non-metals

Group 14-16

Not able to conduct heat or electricity

Brittle

Include gases like oxygen, nitrogen, neon, and chlorine.

Halogens

Group 17

Seven electrons in outer shell

Non-variable oxidation states

Noble Gases

Group 18

Inert gases

Maximum number of electrons in outer shell

Trends in the Periodic Table

One of the most important uses of the periodic table is the tracking of trends throughout the elements. Because the periodic table is set up so that elements are arranged in groups and periods, we can see how the properties of elements change based on where a group is located in the table or based on which period an element is a member of. We can also see how different groups are related to one another in terms of their physical and chemical properties. The most important trends in the periodic table are ionization energy, electron affinity, electronegativity, and atomic radius.

Ionization Energy – As you move left to right across the periodic table, the ionization energy increases. Ionization energy is the amount of energy necessary to pull one electron off of an atom.

Electron Affinity – As you move down a group in the periodic table, the electron affinity decreases. Electron affinity is the energy change that occurs when an atom picks up an extra electron.

Electronegativity – Electronegativity increases as you move left to right across the periodic table. Electronegativity is energy an atom needs to pull electrons away from other elements it was bonded to.

Atomic radius – The atomic radius decreases as you move left to right across the periodic table.

Concept Reinforcement

1. What is an element?

2. Which family of elements explodes in water?

3. Which family of elements has the maximum number of electrons in its outer shell?

4. What trends decrease as you move left to right across the periodic table?

5. Why is the periodic table important in chemistry?

Section 1.5 – Molecules & Compounds

Section Objective

- Describe molecules and compounds

Molecules

Elements combined together chemically are called molecules. **Molecules** are the smallest amount of a substance that still has the properties of that substance. For instance, a single atom of the element of oxygen is expressed by the symbol "O." If you combine two atoms of oxygen, you get a molecule of oxygen, or O_2. This is a constituent of the air that we breathe. This O_2 molecule is the smallest amount of oxygen that you can have.

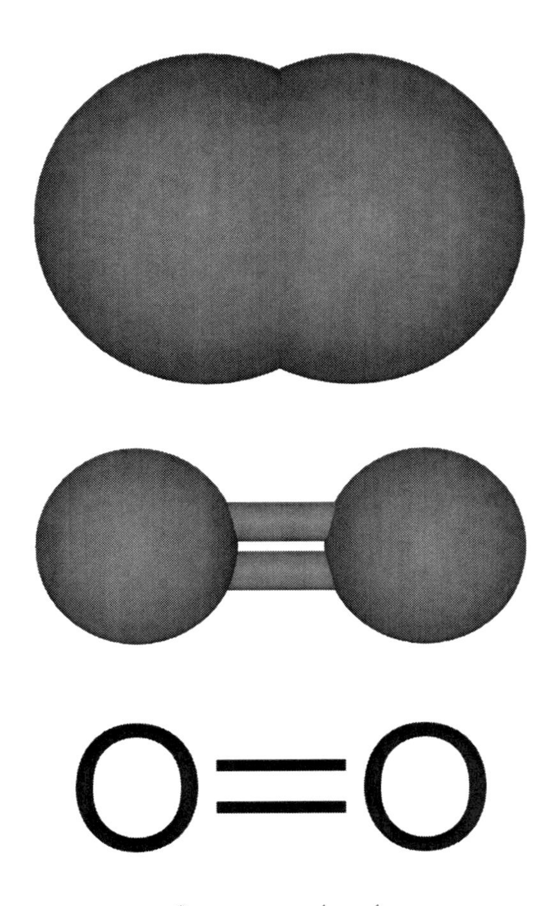

Oxygen molecule

Some molecules are made up of just a single atom. Gold is an element that can exist in its pure state as just one atom. A gold atom is the same as a gold molecule. Other elements like to form very large molecules in their pure form. Phosphorus likes to form a molecule of four phosphorus atoms in its natural state. The smallest amount of phosphorus possible is a molecule of phosphorus, written as P_4.

Compounds and Mixtures

When one atom of oxygen is combined chemically with two atoms of Hydrogen they form H_2O, also known as water. Water is an example of a compound, which is a special type of molecule. A **compound** is a molecule that is made up of two different elements that are chemically bonded together. This causes the new compound to have different physical and chemical characteristics from the original elements. Hydrogen and oxygen are both gases, while water is a liquid. Another example of a compound is rust. Iron is a shiny, hard metal and oxygen is an invisible gas. When these elements combine, they form rust (Fe_2O_3). Rust is a red, brittle solid that is much weaker than pure iron. Oxygen gas, gold, and phosphorus are molecules that are not compounds because they are only made up of one type of atom. These molecules are elements, not compounds.

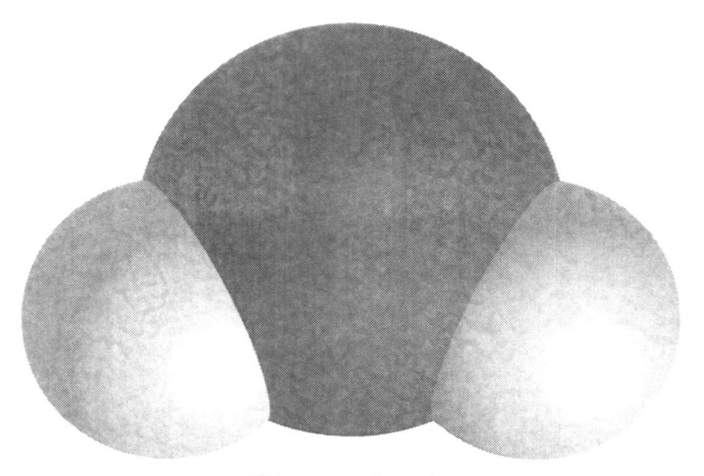

Water molecule

A **mixture** is a combination of two different compounds through mechanical means. The original compounds do not chemically bind, so therefore retain their original chemical and physical properties. Think of combining dry rice and dry beans. You can mix the two things together, but they can always be separated later through mechanical means.

Concept Reinforcement

1. How are compounds different from molecules?
2. How are mixtures different from compounds?
3. Is hydrogen (H_2) a compound, molecule, or mixture?
4. Is carbon dioxide (CO_2) a compound, molecule, or mixture?
5. If you combine water and sand, do you have a compound or mixture?

Section 1.6 – Ionic Bonding

Section Objective

- Discuss ionic bonding

Bonding

Bonding, more specifically known as **chemical bonding**, is the attractive interaction between atoms. Bonding is the way that atoms and molecules join together to form larger compounds. Another way to think of bonding is the force that causes atoms to stick together to form molecules. There are several different types of chemical bonds found in nature, including ionic bonds, covalent bonds, hydrogen bonds, and metallic bonds.

Ionic bonding

Ionic bonding is a type of chemical bonding that occurs between a metal and a non-metal ion. So, what is an ion? An **ion** is an atom or molecule that has either gained or lost electrons. This gain or loss gives the atom a positive or negative electrical charge. Atoms that lose electrons become positively charged ions while atoms that gain electrons become negatively charged ions. These opposite charges cause attractions between the positive and negative ions. The attractions cause the ions to pull together forming a strong bond. This bond is an **ionic bond**.

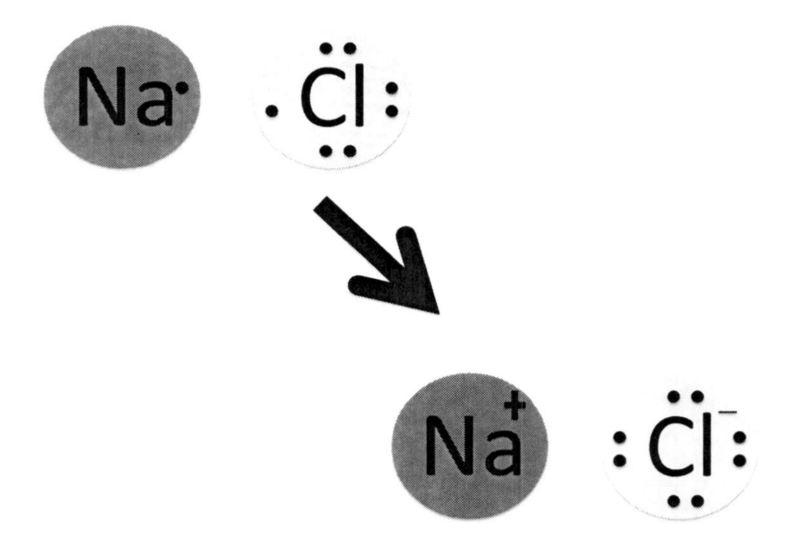

Why does this happen? In the periodic table, we know that electronegativity, as a trend, increases as you move across the periodic table. Electronegativity is the ability of an atom to pull electrons away from another atom.

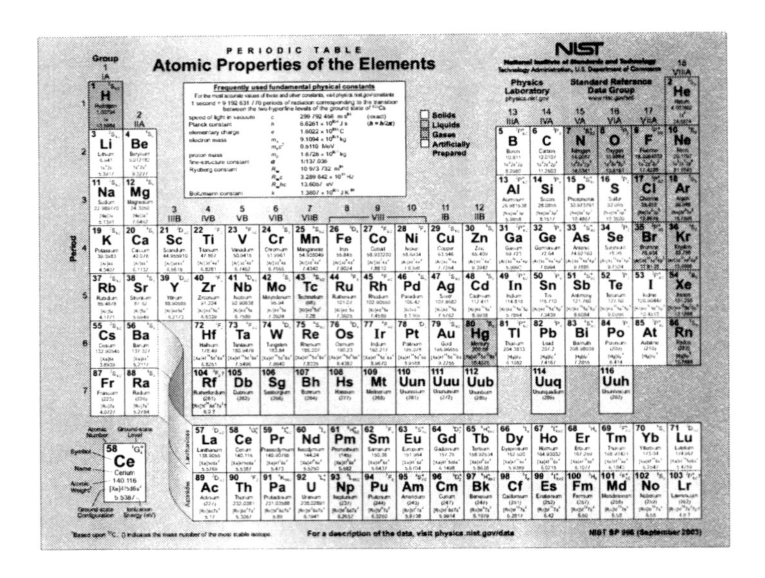

Let's look at simple table salt as an example of an ionic bond. Sodium (Na), a metal gives an electron to chlorine (Cl), a non-metal halogen. Therefore, sodium, which is in Group 1, has a low electronegativity. Chlorine, which is in Group 17, has a very high electronegativity. By combining the two elements in equal amounts, the sodium will donate an electron to the chlorine. This donation creates a positive ion of sodium, and a negative ion of chlorine, which are then joined to create sodium chloride (NaCl).

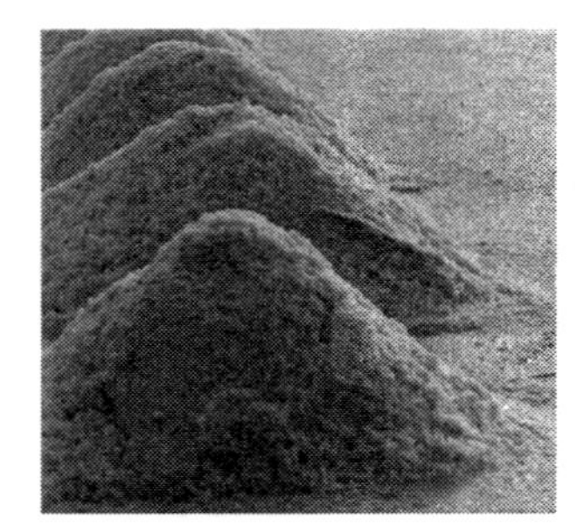

Salt

Concept Reinforcement

1. What is a chemical bond?
2. What is an ion?
3. In ionic bonding are the electrons shared or donated?
4. Lithium and fluorine form an ionic bond and form the compound lithium fluoride (LiF).
    a. Which element has a higher electronegativity?
    b. Which element donates an electron?
    c. Which element forms a negative ion?

Section 1.7 – Covalent Bonding

Section Objective

- Discuss covalent bonding

Bonding

Chemical bonding is the force of attraction between atoms. Chemical bonding allows atoms to join or "stick" together to form larger molecules. There are several different types of chemical bonds found in nature, including ionic bonds, covalent bonds, hydrogen bonds, and metallic bonds. **Ionic bonds** are bonds that form when an atom donates an electron to another atom, forming positive and negatively charged ions. Ionic bonds are formed between atoms with very different electronegativities that are located on opposite sides of the periodic table.

Covalent Bonding

Covalent bonding is a type of chemical bonding where atoms share electrons with each other. This is different from ionic bonding, because electrons are shared with each other as opposed to being donated to form positive and negative ions. Covalent bonds form between atoms with similar electronegativities, meaning that the elements that form covalent bonds are often found close to one another in the periodic table. Covalent bonds are most commonly formed between non-metals, metalloids, and metals in groups 3 through 15 of the periodic table.

Bond Strength

Covalent bonds are considered strong bonds, and are usually stronger than ionic bonds. The strength a covalent bond varies based on a number of factors including which elements are bonded and how many bonds are formed between the atoms. In general, the more electrons that two atoms share, the stronger the bond will be between the atoms

Single Bonds – Single bonds are the weakest type of covalent bond, and are formed from a single pair of electrons being shared. This single pair of shared electrons is shown as a single line between the elemental symbols

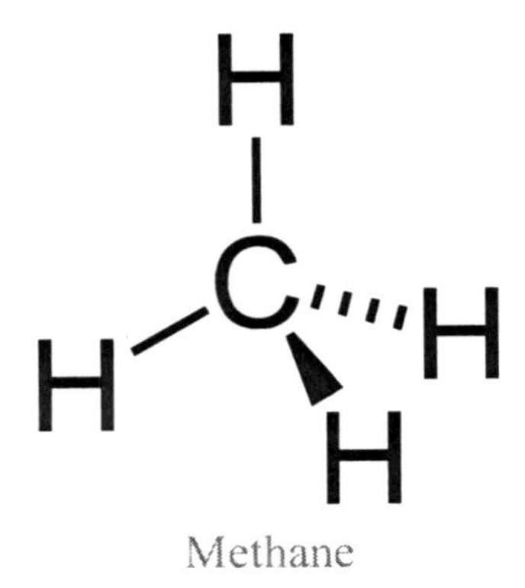

Methane

In the diagram of methane, a single pair of electrons is shared between each carbon and each of the four hydrogen atoms. Carbon is able to form four covalent bonds, so there are four single bonds formed between the carbon and each of the hydrogens.

Double Bonds – Double bonds are formed when two pairs of electrons are shared between elements or molecules. These bonds are stronger than single bonds, and are shown by a double line between both elements.

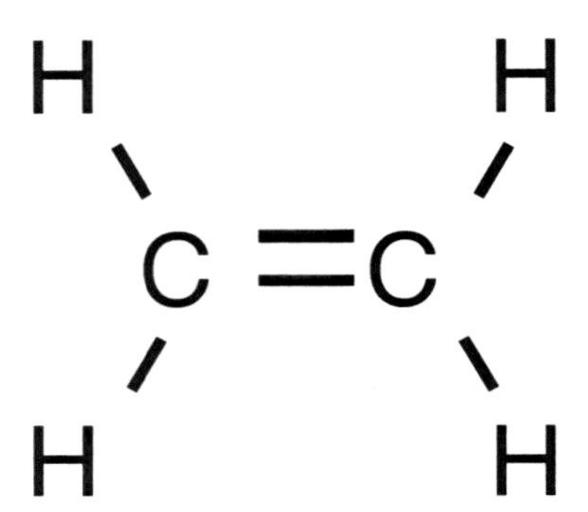

An ethylene molecule

Ethylene is a molecule composed of two methane molecules. Since two pairs of electrons are being shared in this compound, you lose one hydrogen atom when the double bond is created.

Triple Bonds – Triple Bonds are formed when three electrons are shared between elements or molecules. These bonds are even stronger than double bonds, and are shown as three lines between the elemental symbols.

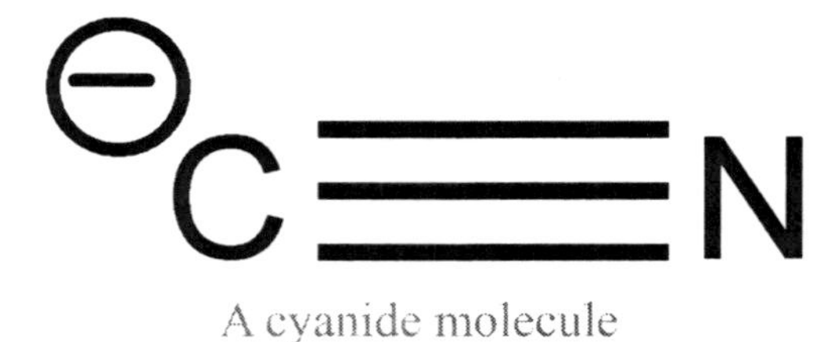

A cyanide molecule

Cyanide is formed when a triple bond is formed between a carbon and nitrogen molecule. As compared to the single bond methane molecule, the carbon had to give up two hydrogen atoms to form the triple bond.

Multiple bonds – There are also less common multiple bonds of four, five and even six bonds. These are commonly found in larger metallic and transition metal compounds.

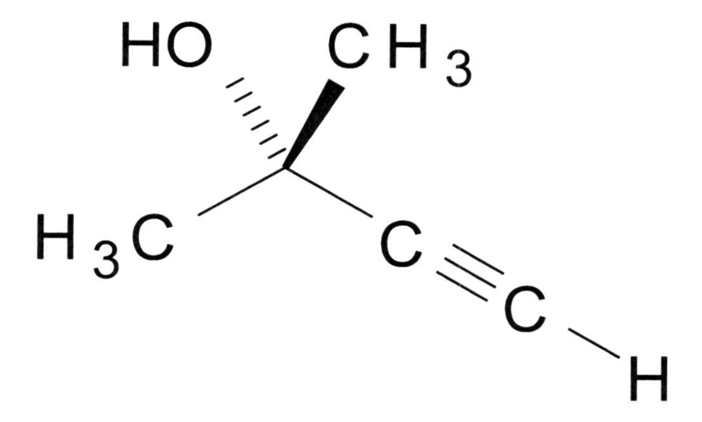

An image of a molecule of 2-Methylbut-3-yn-2-ol

Concept Reinforcement

1. How are covalent bonds different from ionic bonds?
2. How is electronegativity related to the formation of covalent bonds?
3. Which is stronger: a single bond or triple bond?
4. Which is stronger: an ionic bond or a covalent bond?

Section 1.8 – Moles

Section Objective

- Define moles and discuss how they are used

The Mole

To most people a mole is a little furry animal that crawls underground, a blemish on the skin or a spy who hides within an organization stealing secrets. In chemical engineering, though, a mole is even more important. The **mole** (abbreviated as mol) is the basic unit of measurement to determine the amount of a substance. To standardize the measurement, everything is based on the carbon 12 molecule. One mole is equal to $6.02 \cdot 10^{23}$ particles of a substance. This can be atoms, molecules, or any other elementary particle. The number of particles, $6.02 \cdot 10^{23}$, is also known as **Avogadro's constant** and is based on the number of atoms in 12 grams of Carbon 12.

The mole is used to standardize the number of molecules that are used in a chemical reaction. Moles are used to count the number of molecules, not the mass or volume of substance, so that the correct ratio of reactants can be used in a chemical reaction. One mole of any element or compound contains the same number of particles as a mole of any other element or compound. Therefore, one mole of lead has the same number of atoms as one mole of gold. One mole of gold has as many atoms as a mole of carbon dioxide has molecules.

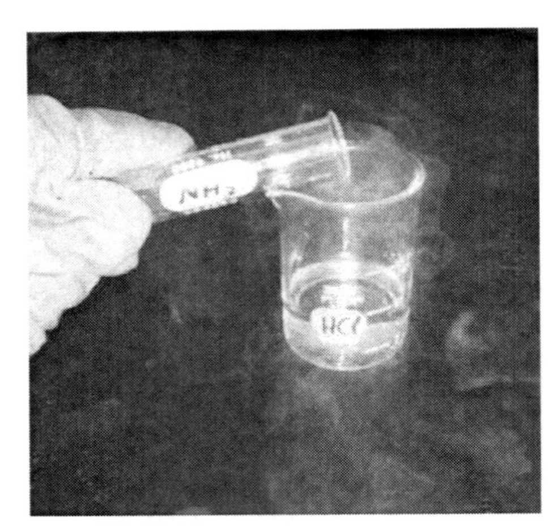

Chemists measure the amount of reactants for chemical reactions in moles.

The standardization of the mole is important to chemical engineering because we need to know the exact quantities of substances if we are going to design a system for them. If we want to build a system to clean up water pollution, we need to be able to accurately measure the contaminants to design a system to remove them efficiently.

Because atoms and molecules are too small to be seen with the naked eye and $6.02 \cdot 10^{23}$ is too large of a number to count to, we need an easier way to measure moles. Moles can be measured using the molar mass of an element. The **molar mass** is the mass of one mole of an element. The molar mass is determined by multiplying the atomic weight of an element by the **molar mass constant** of 1 g/mol. The **atomic weight** is the ratio of the average mass of an element to 1/12 the mass of an atom of Carbon 12. The atomic weight of each element is listed under the element's symbol in the periodic table.

Example 1: Find the molar mass of helium.

To find the molar mass of helium, we must first find the atomic weight of helium. Using the periodic table, we can see that helium has an atomic weight of 4.002602, which we can round to 4 for our purposes. To find the molar mass, we simply multiply the atomic weight by the molar mass constant of 1 g/mol.

$4 \cdot 1 \text{ g/mol} = 4 \text{ g/mol}$

Answer: The molar mass of helium is 4 g/mol.

Example 2: Find the molar mass of gold.

To find the molar mass of gold, we must first find the atomic weight of gold. Using the periodic table, we can see that gold has an atomic weight of 196.966569, which we can round to 197 for our purposes. To find the molar mass, we simply multiply the atomic weight by the molar mass constant of 1 g/mol.

$197 \cdot 1 \text{ g/mol} = 197 \text{ g/mol}$

Answer: The molar mass of gold is 197 g/mol.

So, we now know that if we have 4 grams of helium, we have 1 mole of helium. If we have a mole of gold, it would have a mass of 197 grams.

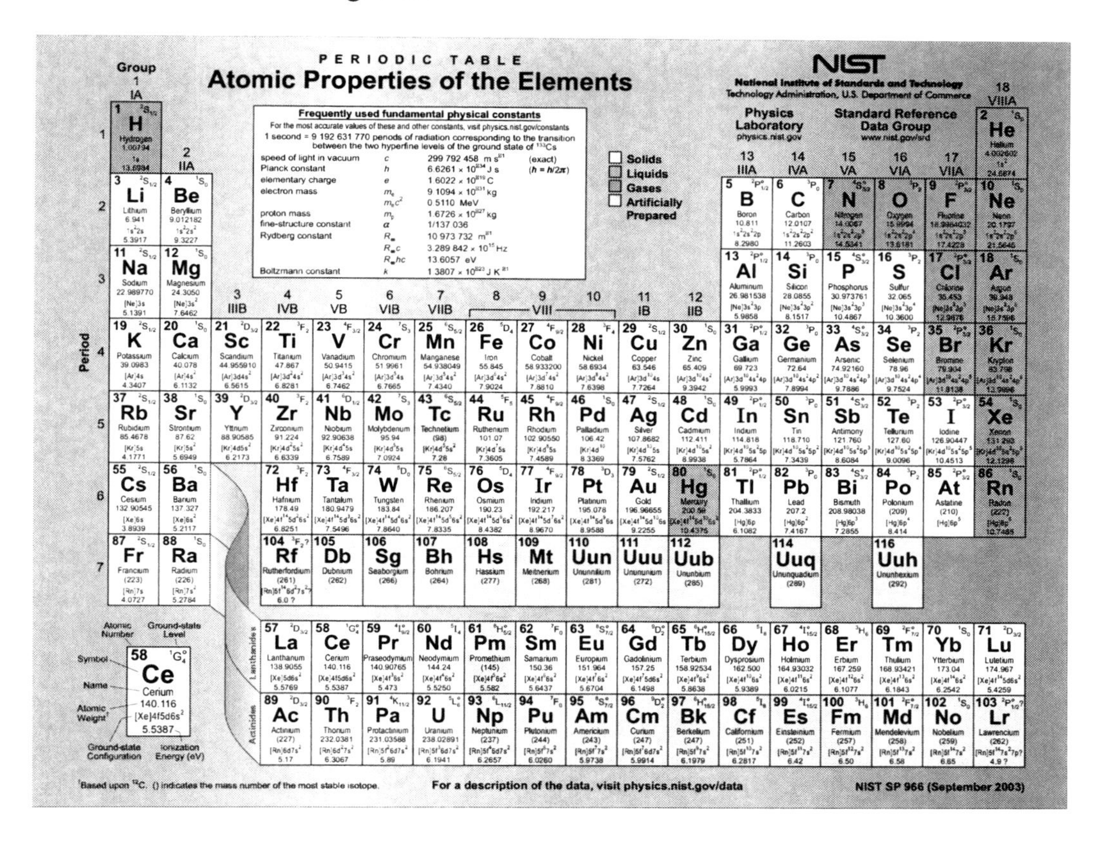

Concept Reinforcement

1. Which element is used as the basis for the mole?
2. How many atoms are found in 1 mole of iron?
3. What is the atomic weight of silver?
4. What is the molar mass of lithium?

Section 1.9 – Density, Specific Gravity & Flow Rate

Section Objective

- Describe density, specific gravity, and flow rate

Density

Density is a measurement defined as the mass per unit volume under specified conditions of temperature and pressure. Another way to think about density is how closely or tightly packed the matter is within an object. The formula for finding the density of an object is:

$\rho = m / V$

where

ρ is the density of the object

m is the mass of the object in grams

V is the volume of the object in milliliters

Example: A small metal plate weighing 5,000 g has a volume of 25 mL. Find the density of the object.

To find the density of the object, we use the formula for density.

$\rho = m / V$

ρ = The density of the metal plate

m = 5,000 g

V = 25 mL

ρ = 5,000/25 = 200 g/mL

The density of the metal plate is 200 g/mL

Density is the same for any substance, regardless of the amount of the substance present. For example, gold has the same density whether it is a small speck of gold or a large gold bar. As the size and volume of the piece of gold increases, the mass increases at the same rate, so the density is the same. The same is true for compounds like water – a small drop of water has the same density as a gallon of water. In the example above, if a larger object was made out of the same metal as the small plate, it would still have a density of 200 g/mL.

Why is this important to a chemical engineer? The first use of density is credited to the Greek scientist Archimedes. Archimedes was working on a problem to determine if a substance other than gold was being used to make a gold crown. So how could Archimedes determine this without destroying the crown? History has it that Archimedes came up with the idea while in the bath tub. He realized that different substances have different densities would displace different amounts of water. So by placing the crown, and an equal amount of gold in a container of water, the crown if made of all gold, should displace the same amount of water.

Oil painting of Archimedes by the Sicilian artist Giuseppe Patania (1780-1852)

So how can density be used in engineering? Density can be used in such endeavors as ship building. Wood for example is less dense than water, therefore wood floats on water. Calculating the density of a substance can help you determine the purity of the substance along with countless other applications.

Specific Gravity

Specific gravity is defined as the ratio between density and the weight of water at a specific temperature and pressure. If a substance has a specific gravity greater than 1, the substance will sink when placed in water. If the specific gravity is less than 1, the item will float in the water. The formula for finding the specific gravity of an object is:

$SG = \rho_{object} / \rho_{water}$

where

SG is the specific gravity of the object

ρ_{object} is the density of the object

ρ_{water} is the density of water

The density of water varies based on its temperature. Warmer water is less dense than colder water, until the water reaches 4 °C. At temperatures below this point, water starts to become less dense as it freezes. This explains why ice cubes always float to the top of a glass of water – they are less dense than the liquid water. The density of water at its densest point (4 °C) is used as the standard for specific gravity. The standard used for ρ_{water} is 1 g/mL or 1,000 kg/m^3. Engineers may occasionally use other values for the density of water depending on what conditions they are working with in a project.

Example: A carbon fiber that is being developed in a chemical plant has a density of 740 kg/m^3. What is the specific gravity of the plastic?

To solve this problem, we divide the density of the plastic by the density of water.

$SG = \rho_{carbon\ fiber} / \rho_{water}$

SG = the specific gravity of the carbon fiber

$\rho_{carbon\ fiber}$ = 740 kg/m^3

ρ_{water} = 1,000 kg/m^3

$SG = 740 / 1{,}000 = 0.740$

Answer: The specific gravity of the carbon fiber is 0.740. Because this value is less than 1, it tells us that the carbon fiber will float when placed in water.

Flow Rate

Flow rate is the volume of fluid that moves over a certain area at a specified amount of time, like gallons per hour (US Standard) or cubic meters per second (metric). Flow rate is used to calculate volumes through systems like water plants, or how much gas moves through a furnace. These calculations are all important to an engineer in building and refining designs. The formula for finding the flow rate of a substance in a pipe is:

$Q = A \cdot v$

where

Q is the flow rate through the pipe

A is the cross section area of the pipe in square meters (m^2)

v is the velocity of the substance moving through the pipe in meters per second (m/s)

Flow rate is usually measured in cubic meters per second (m^3/s)

Example: A pipe with a cross section area of 0.25 m^2 has water flowing through it at 150 m/s. What is the flow rate of the pipe?

To solve this problem, we need to multiply the cross section area of the pipe by the velocity of the water.

$Q = A \cdot v$

Q = flow rate of the pipe

$A = 0.25\ m^2$

$v = 150$ m/s

$Q = 0.25\ m^2 \cdot 150\ m/s = 37.5\ m^3/s$

Answer: The flow rate of the pipe is 37.5 m^3/s.

Concept Reinforcement

1. Which would have a greater density: a granite pebble or a granite boulder?
2. Find the density of an object with a volume of 200 mL and a mass of 8,000 g.
3. A piece of metal has a mass of 5,000 g and a volume of 250 mL. Find the specific gravity of the object and decide if it will sink or float when place in water.
4. Find the flow rate through a pipe that has water flowing through it at 120 m/s and has a diameter of 0.5 m^2.

Section 1.10 – Concentration

Section Objective

- Define concentration and its importance in chemical engineering

Solutions

A **solution** is a type of mixture in which a substance called the **solute** is dissolved in another substance called the **solvent**. When a solution is made, the solute becomes evenly distributed in the solvent. Solutions are a special type of mixture, but are often appear to be pure substances due to the even distribution of the solute in the solvent.

When you open add sugar to a cup of hot coffee, the sugar appears to disappear. Where does the sugar go? The sugar, which is the solute, is dissolved into the coffee, which is the solvent. When the sugar disappears, it becomes evenly distributed in the coffee so that the coffee is equally sweet throughout.

Another example of a mixture is salt water. If you pour salt into water and stir the salt, it will eventually disappear. The salt, which is the solute, becomes dissolved in the water, which is the solvent. How do we know that the salt is still in the water? First, the water will now taste salty. Second, if we leave the salt water sitting out for several days, the water will evaporate and leave a crusty layer of salt around the edges of the glass.

Concentration

In its simplest form, **concentration** is the amount of one substance in another substance. This is a critical measurement of chemical engineering since the concentration of a substance can determine the rate of a reaction or even the amount of a contaminant. To find the concentration of a solvent in a solution, we use the following formula:

Concentration = Amount of solute / Amount of solvent

Concentration can be quantitatively expressed in a variety of ways.

Percentage – a substance can be 34.2% of a solution or mixture. This means that out of a possible 100%, this substance is 34.2%, while the remaining 65.2% is made up of other constituents.

Parts per – For low concentrations the parts per (pp) measurement is used. Concentrations may be expressed as parts per million (ppm) or parts per billion (ppb). Because a billion is 1,000 times smaller than a million, 1 ppm is the same as 1,000 ppb. Parts per billion is used to measure incredibly tiny concentrations.

Example: A chemical reaction requires 100 ppb of benzene to be dissolved in water. How much benzene is needed as expressed in ppm?

Because a billion is 1,000 times smaller than a million, 1 ppm is the same as 1,000 ppb. Therefore, we will divide the measurement in ppb by 1,000 to find the same measurement in ppm.

100 ppb / 1,000 = 0.1 ppm of benzene in water

Mass to Volume – In this form of measurement, we take mass, such as grams, and relate it to a volume, such as milliliters or cubic meters. Mass to volume ratios are often measured in micrograms per liter (µg/L) or milligrams per liter (mg/L).

Example: A chemical engineer finds that there are 50 mg of hydrochloric acid (HCl) in 10 liters of water taken from a stream that runs nears a chemical plant. Find the concentration of HCl in the stream in mg/L.

To find the answer, we simply divide the mass of the HCl by the volume of water.

50 mg HCl / 10 L water = 5 mg/L

Answer: The concentration of HCl in the stream is 5 mg/L.

Molarity–The most common quantitative way of expressing concentration is molarity. Molarity is measured as the number of moles of a substance in one liter of solution. Although this form, along with mass to volume can be very precise, issues such as thermal expansion can affect the precision of the measurement. The formula for finding the molarity of a substance is:

Molarity = Moles of solute / Liters of solution

The unit used for measuring molarity is moles per liter or mol/L.

Example: 15 mol of sodium hydroxide (NaOH) is added to distilled water to make 100 liters of solution. What is the molarity of the solution?

To find the molarity, we must divide the moles of solute (NaOH) by the volume of solution in liters.

Molarity = Moles of solute / Liters of solution

Molarity = 15 mol NaOH / 100 L of solution = 0.15 mol/L

Answer: The molarity of the solution is 0.15 mol/L

Molality – Molality is defined as the number of moles of a solute divided by the number of kilograms (kg) of the solution.

Other forms of concentration measurements can be used. In engineering, the use of various types of systems allow for the use of varied measurements. Through standardizing your measurement, you can easily move from one type of unit to another.

Importance of Concentration in Chemical Engineering

Air pollution

Chemical engineers often use concentrations to measure the amount of a solute present in a solution. Knowing the amount of solute is very important when adding reactants to a chemical reaction – adding too much of a solute would be wasteful and adding too little would waste other chemicals used in the reaction. Efficiency is a very important concept in engineering to save production costs and reduce pollution.

Just like you may buy concentrated juice, detergent, or bleach at the supermarket, most products are shipped in a concentrated form. Most liquid chemicals are shipped around the world as concentrated solutions to reduce the weight and size of the chemical during shipping. These concentrated chemicals are almost never used in their concentrated form because they are usually too strong in this form. Another factor is that a diluted chemical is more forgiving if there is a small error in measurement of how much chemical to add. A few extra drops of concentrated acid could make a big difference in a reaction compared to a few extra drops of a diluted solution of the acid.

A final use of concentration is engineering is in relation to the environment. Chemical plants are common targets for environmental testing and regulation because of the threat of pollution. Many of the chemicals used in the plant as well as waste products are very dangerous if allowed to escape into the water, soil, or air near a chemical plant. Chemical engineers are often responsible for the testing of the environment surrounding a chemical plant to ensure that pollution is not occurring. Chemical plants that allow high levels of pollution to escape face huge fines as well as the possibility of being shut down by the government. Chemical engineers checking for pollution often measure the concentration of chemicals in parts per million and parts per billion because environmental laws are written using these measurements of concentration. Most environmental laws allow for a very small amount of chemicals to be present in the air, soil, or water, so engineers must know if they are exceeding their legal limits for pollution or not.

Concept Reinforcement

1. 5 ppm is equal to how many ppb?
2. 500 ppb is equal to how many ppm?
3. A chemical engineer needs to have 25 kg of potassium chloride (KCl) dissolved in 250 liters of water to complete a chemical process. Find the concentration of KCl needed in kg/L.
4. 350 mol of sodium fluoride (NaF) is added to distilled water to make 500 liters of solution. What is the molarity of the solution?

Section 1.11 – Analysis of Multi-Component Solutions and Mixtures

Section Objective

- Explain how to analyze multi-component solutions and mixtures

Chemical Analysis

Chemical analysis is the determination of the chemical make up of a mixture or solution. This determination is performed by a variety of chemical and physical means. In these determinations you are trying to find out what is present (qualitative) and/or how much is present (quantitative).

Qualitative Analysis

Qualitative analysis is a scheme of analysis to determine the presence of an element, inorganic compound, organic compound, or functional group. This scheme usually performed in an aqueous matrix, tests the unknown substance through physical means such as boiling point, flash point, or reaction by the addition of known chemicals. A positive reaction identifies the specific element, compound or functional group.

Quantitative Analysis

Traditional Methods

Quantitative analysis is used to determine how much of a substance is present in a sample. Traditional methods for quantitative analysis are a more generalized method of analysis for the determination of solutions and mixtures. These are the classic techniques that Hollywood and the general public have had about chemistry. The scenes of bubbling flasks and long columns running across the lab are a somewhat exaggerated view of these classic techniques.

Laboratory of Experimental Medicine and Cancer Research of Johnston Laboratories at the University of Liverpool, Liverpool, England, in the United Kingdom. This photograph dates from the formal opening of the laboratories in 1903.

Gravimetric – Gravimetric analysis involves the determination of the mixture through the change in weight of the substance. A substance is weighed initially, and a chemical is added to the substance to cause a chemical reaction. After the reaction, the substance is reweighed to determine the amount of substance present.

Titrimetric – Titrimetric analysis, also known as **titration**, involves the addition of a reactant to a solution that an indicator chemical has been added too. The indicator chemical changes color when the solution has completely reacted with the reactant being added. Additional titration techniques can involve potentiometric techniques such as pH changes.

Instrumental Techniques

Instrumental techniques were primarily developed in the twentieth century, with the last couple of decades showing the most growth. By moving these classic techniques to instruments, and expanding their potential, chemists were able to improve the quality of their measurements. The reason for this is that instrument perform a procedure the same way, every time, therefore, measurements are more reproducible. In addition, since the measurements are reproducible, the precision of the measurements can be improved over time.

Spectroscopy – Spectroscopy is the measurement of a substance based on the wavelength of light that they absorb or emit. Some common types are:

Ultraviolet/Visible spectroscopy (UV/VIS) – In this version, substances is analyzed by light in the ultra-violet or visible range of the spectrum. An indicator is added to a solution, developing a color. The solution is then analyzed for the color that matched that spectrum. Based on the intensity of the color, the quantity can be determined.

Atomic absorption spectroscopy (AA) – In this style of spectroscopy, the substance is introduced into a flame of a known energy and analyzed at a certain wavelength. The difference between the known energy of the flame and what is recorded at a detector is based what is proportionally absorbed by the substance.

Atomic emission spectroscopy – This type of spectroscopy works in a similar manner to atomic absorption, except that it measures the wavelength of light that is emitted from excited substances.

Infrared Spectroscopy (IR) – This type of spectroscopy deals with the infrared portion of the spectrum. It can be used to determine structures and therefore composition of substances.

Other types of spectroscopy are: Raman spectroscopy, x-ray fluorescence spectroscopy (XRF), nuclear magnetic resonance spectroscopy (NMR) as well as others currently being developed.

Separation Techniques

There are two main types of separation techniques, electrophoresis and chromatography. In all separation techniques, you employ a repeated separation technique that allows you to get the same results on the same substance every time you perform the analysis.

Electrophoresis – In electrophoresis an electric field is applied to a gel. The substance to be analyzed is added to the gel, and migrates in bands based on the type of gel, the size of the molecules, and the charge of the molecules. Large molecules have a very difficult time moving through the pores in the gel, and therefore move very slowly. Small molecules easily fit through the pores in the gel, and move quickly through the gel. After several hours have passed, the mixture has been separated into its components based on molecular size. Electrophoresis is most commonly used for the separation of long chain molecules like DNA, RNA, and proteins.

Chromatography – In the classic sense, chromatography was performed by filling a column with some solid substance, and pouring an unknown substance through. Based on what the solid is, and the substance, the individual components of the substance separate as they move through the column. In current technology, there are two main forms of chromatography, gas, and liquid chromatography. Both use a column to separate the analytes of interest, along with a detector after the column to identify and or quantify the analyte. In chromatography, an analyte is identified by what is called retention time. That is the time from when the analyte enters the column, until it is recorded at the detector.

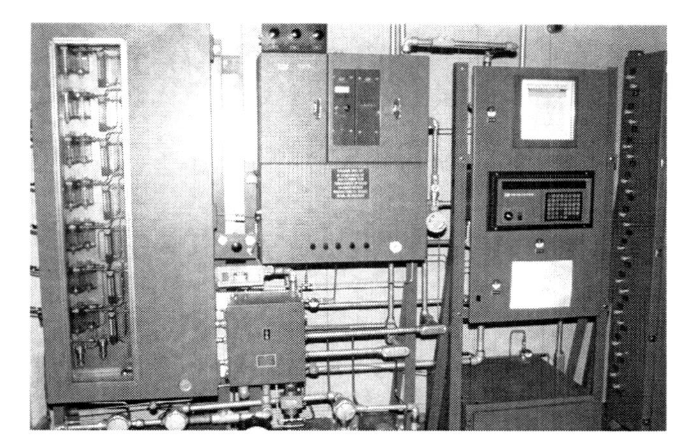

Gas chromatograph

Gas chromatography (GC) – As the name suggests, gas chromatography involves the separation of a gas into its constituent parts. In the original GC columns, there was a very basic packing material in the column that caused the compounds to slow down in their movement through the column to the detector. As technology has advanced, columns have changed. In order for columns to perform the separations, they have gotten longer and thinner. The first columns were less than one foot long with a diameter of ¼ of an inch. Now columns can be over 100 meters long with diameters as small as 0.1 micrometers. These are generally called capillary columns. Instead of being packed, they are now coated with various polymers to add in the separation. The identification of the substance is performed by using a detector. A detector is a device that gives a response when an analyte passes through it. In addition, since the response can be just a tiny change in an electric current and amplifier is used to make the detection apparent. Now detectors can be a non-specific or something very specific. The more specific the detector, the more sensitive the detection can be. Here is a list of some of the more popular detectors:

Flame Ionization Detector (FID) – An FID is a non-specific detector that is a basically a hydrogen flame. So any compound that burns in a flame will easily be detected by an FID. Think of how gasoline as opposed to salt water acts in a flame. The constituents in gasoline would give a huge response to an FID, where salt water does not burn, and therefore would not be detected.

Nitrogen-Phosphorous Detector (NPD) – As the name implies, this detector is sensitive to compounds that have nitrogen and phosphorous in them. The more nitrogen and/or phosphorous in the analyte, the higher the response on the detector.

Photo-ionization Detector – (PID) This detector can detect the unsaturated bonds in organic compounds, so a compound like benzene which is aromatic, would be detected, where a straight chain hydrocarbon like hexane would not be detected.

Some other types of detectors:

Electron Capture Detector (ECD) – detects halogens

Flame Photometric Detector (FPD) – Detects sulfur compounds

Liquid Chromatography – As the name suggests involves a liquid being separated into its constituent parts. In liquid chromatography a liquid is introduced into a column. Here the substance remains a liquid, and is separated as in GC so that it can be detected once it leaves the column. LC columns tend to be shorter than GC columns. In general, they are only a few inches long, and can be either packed or capillary type. As in GC, there are various detectors that are used. Some of the more popular types are:

Ultra-violet/Visible Detector (UV/VIS) – This detector identifies compounds that absorb light at a specific wavelength.

Fluorescence Detector – This detector identifies compounds that fluoresce at certain wavelengths.

Specialized Detectors

Mass Spectrometry–Mass spectrometry is a technique that identifies substances by their mass to charge ratio by using either electric or magnetic fields. In this technique, an analyte is ionized by electric or chemical means primarily, and a fragment of the original compound is created. This fragment has a mass to charge ratio. If this procedure is performed the same way each time, you will get the same fragment pattern each time. This is called a compound's mass spectrum. When this technique is coupled to chromatography, a hyphenated technique such as GC-MS or LC-MS is formed. In this hyphenated technique there is the retention time which identifies the analyte, and its mass spectrum which confirms the identification. The technique is very powerful in identifying unknown chemicals.

Concept Reinforcement

1. How is qualitative analysis different from quantitative analysis?
2. What technique uses weight for chemical analysis?
3. How would you separate the various components in gasoline?
4. Which technique is used to separate long chain molecules like DNA?

Section 1.12 – Choosing a Basis

Section Objective

- Explain how to choose a basis

What is a Basis?

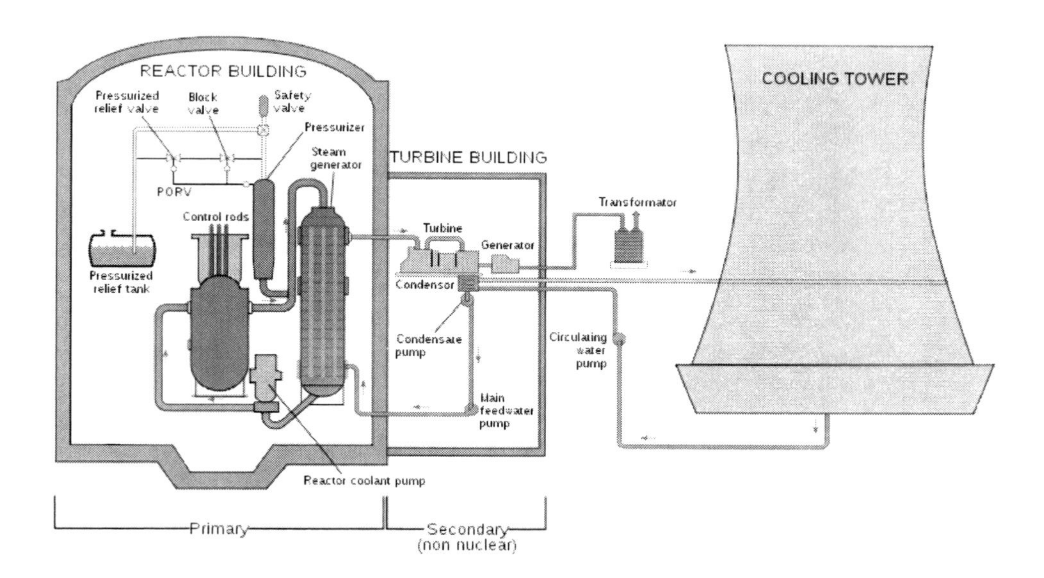

A **basis** is a starting point, a foundation, or the basics that you need to start or perform a project. In chemical engineering, the basis is the plan used to design a chemical plant or a new process within a chemical plant. The basis is the theoretical part of a chemical process – what we think will be needed to run a chemical plant for many years. The basis is often changed and revised over time and presented to zoning boards or planning commissions prior to a chemical plant being built.

Many of the most important decisions that need to be made to successfully complete a project are made during the determination of the project's basis. The basis lays many considerations for a new project, including:

- The units of measurement to be used
- The schematics of the machinery
- The materials needed
- The outputs from the process
- Safety considerations
- Pollution control strategies

- Laws concerning pollution, noise, worker safety, and zoning of property
- Redundant and backup systems
- How much product will be produced per hour, day, etc.
- The estimated cost of the product

Choosing a Basis

In engineering, we need to start at one point, and stick with it. In other words, we need to choose what system and what type of units we are going to work with and not change. This allows for ease of collaboration and review by outside organizations that monitor the work that is done in the chemical plant. The use of a single system of measurement helps to prevent mathematical errors that can occur when units have to be converted from one system to another throughout a production process.

Let's say that we are designing a device to manufacture a specific chemical. We need to know some of the following items:

How much of the final product do we need to make?

At what ratio do we mix the starting chemicals?

How are we going to measure these chemicals?

Working with these parameters, we decide that we are working with weights and mass. Do we work in pounds (lbs) or kilograms (kg) or even milligrams (mg)? These are important to decide early. If we use different units, we will have to perform possibly complicated conversions.

Sometimes chemical engineers have to deal with the local standards when determining their basis. In the United States, chemical suppliers, buyers and local government may use the US Standard system of measurement. In this case, it may be easier to use US Standard measurements throughout the process. In the rest of the world, the metric system is much more commonly used and is likely to be used throughout. But what if a product is made in the United States with components that are imported from another country and shipped overseas? If the inputs and outputs are measured in metric and local regulations are measured in US Standard, the engineers designing the plant must which system of measurement is the most suitable for their needs.

Deciding on the scale of measurement is important, too. Some chemicals are made in very small batches while others are made in large scale productions. If we decide to use metric masses, should the product be measured in milligrams, grams, kilograms, or even metric tons? These units can be converted easily in the metric system, but to keep things simple and prevent mathematical error it is better to use the same units throughout the process.

Concept Reinforcement

1. What is a basis?
2. Why is it important to develop a basis before beginning a production process?
3. What kinds of things does an engineer need to consider when developing a basis?
4. Why is it a bad idea to mix different units of measurement?

Section 1.13 – Temperature

Section Objective

- Define and discuss temperature

What is temperature?

Temperature is something that we are all familiar with. We know if something is hot, or if something is cold. When we measure our own temperature we know that 98.6 °F is the normal temperature for the human body. We know that water boils at 212 °F but at only 100 (°C) degrees on the Celsius scale. But what does temperature really mean?

Temperature is a measurement of the motion of the particles in a substance. As the particles that make up a substance move faster, the substance heats up and its temperature increases. As energy is removed, and the particles slow down, the substance gets colder and its temperature decreases. When all energy is removed from a substance, the system reaches a theoretical position of absolute zero (−273.15 °C or −459.68 °F). That equates to absolutely no movement in a system. This has never been observed, although scientists have approached this point to within less than 1 degree Celsius.

How is Temperature Measured?

In chemical engineering, a device is needed to accurately measure temperature. Such a device is called a **thermometer**. A thermometer is a long slender tube filled with a liquid that expands at an equal rate as it is heated and contracts at a similar equal rate when it cools. We are all very familiar with the two basic versions that are around. One is filled with a silver liquid, which is actually mercury; the other is filled with a red liquid which is alcohol. These two types work equally well in the temperature ranges that humans live and work.

Now that we have the device, we have to come up with the scale, or how we read the temperature. In this there are two issues, what does the actual temperature mean, that is what does something like boiling or freezing water mean in terms of a reading and what is the difference in divisions between something such as boiling a freezing water?

When using a thermometer, it is important to know which temperature scale is being used. Several different temperature scales have been developed over the years, mostly based on absolute zero, the boiling point of water, and the melting point of water.

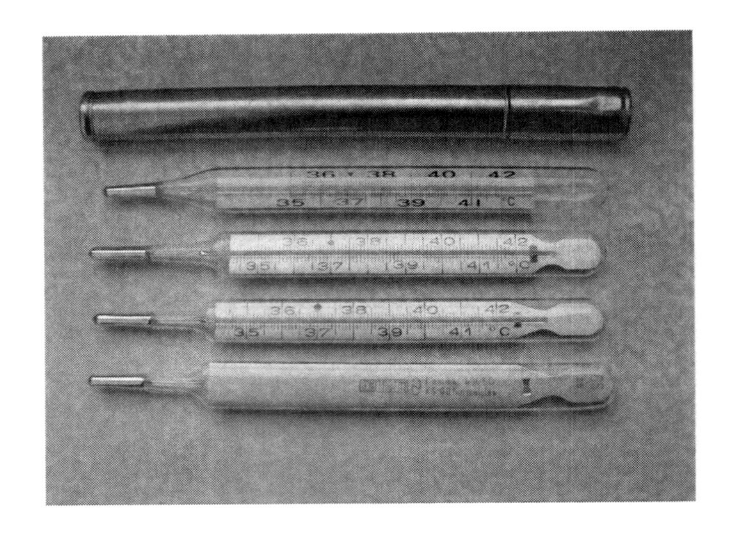

In 1724, Gabriel Fahrenheit came up with one of the first accepted scales for temperature, which was the Fahrenheit scale. This is the scale that is used throughout most of the non-scientific population in America. In this scale, he measured the freezing point in water and marked it as 32 degrees. He then measured the boiling point of water and marked it as 212 degrees. This gave him what he thought was a nice round number of 180 degree difference between the two points. Since that time many other temperature scales have come into use. The most popular is the Celsius scale, which is used by most of the world along with majority of the scientific community. In this scale, water freezes at 0 degrees, and boils at 100 degrees.

Other temperature scales used in science and engineering are:

Scale	Freezing Point	Boiling Point
Kelvin	272.15	373.13
Rankine	291.67	671.64
Delisle	150	0
Newton	0	33
Reaumeur	0	80

Converting Between Temperature Scales

Engineers often need to make conversions between temperature scales, particularly between the Fahrenheit and Celsius scales. To convert a temperature between the Celsius and Fahrenheit scales, use the following formulae:

From Fahrenheit to Celsius: $°C = 5/9 \cdot (°F - 32)$

From Celsius to Fahrenheit: $°F = 9/5 \cdot °C + 32$

The Kelvin scale is often used by scientists. The Kelvin scale uses the same units as Celsius but is based on **absolute zero**, the point at which the molecules in a substance have a kinetic energy of 0.

Absolute zero is considered to be −273.15 °C or −459.68 °F. Because the units are the same, a change of 1 degree Celsius is the same as a change of 1 degree Kelvin. The only difference between the two scales is the point that is considered to be 0 in each scale.

To convert a temperature between the Kelvin and Celsius scales, use the following formulae:

From Celsius to Kelvin: K = °C + 273.15

From Kelvin to Celsius: °C = K – 273.15

Example 1: A chemical reaction must take place at 200 °F in order to be most efficient. What is the temperature that the reaction must occur at in degrees Celsius?

To solve this problem, we need to use the formula for converting from Fahrenheit to Celsius:

$°C = 5/9 \cdot (°F - 32)$

$°C = 5/9 \cdot (200 - 32)$

$°C = 93.3333\ldots \approx 93.3\ °C$

Answer: The chemical reaction must take place at 93.3 °C.

Example 2: A chemical reaction must take place at 200 K in order to be most efficient. What is this temperature in Celsius?

To solve this problem, we need to use the formula for converting from Kelvin to Celsius:

$°C = K - 273.15$

$°C = 200 - 273.15$

$°C = -73.15\ °C$

Answer: The chemical reaction must take place at -73.15 °C

Concept Reinforcement

1. What temperature scale is used by most Americans?
2. At what temperature does water boil on the Kelvin scale?
3. A chemical reaction must take place at 50 °C in order to be most efficient. What is this temperature in Kelvin?
4. A chemical reaction must take place at 50 °C in order to be most efficient. What is the temperature that the reaction must occur at in degrees Fahrenheit?

Section 1.14 – Pressure

Section Objectives

- Define and discuss pressure

Pressure

Pressure is a physical quantity that we all understand on a very basic level. When we press our finger into the palm of our hand, we feel some force where the finger touches the palm, as well on the tip of the finger. That is a pressure. Another means of viewing pressure is experienced by scuba divers. As a diver goes deeper, the pressure increases due to the weight of the water on his body increasing. We experience this change in pressure when we change our altitude by flying in an aircraft or by driving from sea level up into the mountains. As a person moves to a higher altitude the air pressure around the person decreases, causing their ears to pop.

Image courtesy of the National Oceanic and Atmospheric Administration

As a physical quantity, **pressure** is defined as force per unit area. Pressure is often measured in pounds per square inch (psi or lb/in^2) in the US Standard System. The psi measurement is often seen when we add air to tires. Pressure is also measured in Pascals (Pa) in the metric system. A Pascal is a force of 1 Newton over an area of 1 square meter (1 N/m^2). The formula for finding pressure is:

$p = F / A$

where

p is the pressure, measured in psi (US) or Pascals (metric)

F is force being applied, measured in pounds (US) or Newtons (metric)

A is the area over which the force is applied, measured in square inches (US) or square meters (metric)

Example: An industrial press has an area of 500 in^2 over which it exerts force to press steel plates. If the press is operated with a force of 10,000 pounds, what is the pressure exerted by the press in psi?

To solve this problem, we use the formula for pressure:

$p = F / A$

p = pressure in psi

F = 10,000 pounds of force

A = 500 in^2 area of the press

P = 10,000 / 500 = 20 psi

Answer: The press exerts 20 psi of pressure.

Measuring Pressure

Pressure is measured in a variety of ways. The most basic way to measure it is by use of a manometer. A **manometer** is a device that has a u-shaped tube filled with water or mercury. Pressure is measured based on the pressure on the column, the liquid moves and the displacement is measured. It is then reported in inches of water or millimeters of mercury (mm Hg). This unit is commonly used in weather studies. Think of changes in pressure when a front moves through. That is measuring pressure changes in the atmosphere.

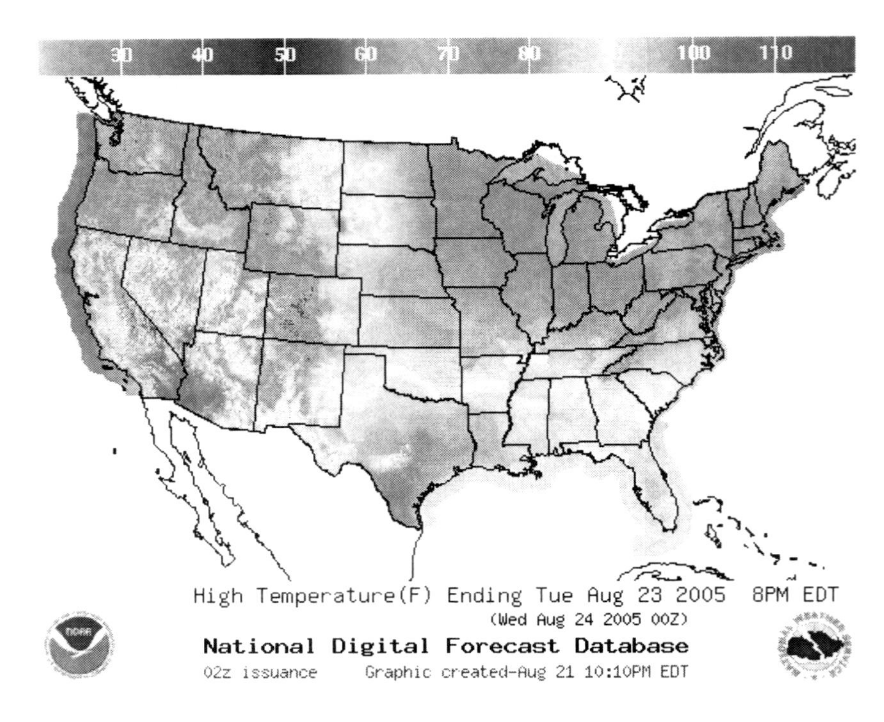

Why is pressure important in engineering?

As one of the basic physical properties, pressure is involved in many basic engineering disciplines. Pressure has applications in making chemicals, as many chemical reactions need to be performed under controlled conditions of pressure and temperature. When designing chemical plants, chemical engineers must take pressure into consideration when planning the strength a pipe needs to be to contain a reaction or how much pressure is needed to make a reaction run properly.

The most common uses of pressure in chemical engineering are in situations dealing with liquids and gases in storage containers or gases produced as a product of a reaction. If the pressure of a liquid or a gas inside of a storage tank increases beyond the ability of a tank to withstand pressure, the tank will explode with disastrous effects. Pressure gauges are installed to monitor the pressure in these tanks and trigger vents to release pressure in the case of an emergency. Many chemical reactions produce gases as a product, and these gases can accumulate until pressures reach unsafe levels. Chemical engineers must plan for the pressures produced by gases by using containers and equipment that can withstand high pressure and by designing overflow tanks or purge valves to remove excess gases from reaction chambers.

Pressure is not always something that chemical engineers try to shy away from. High pressure can be used to turn gases into liquids, like liquid oxygen, hydrogen, helium, and nitrogen. When the pressure is reduced, these liquids turn back into their gas forms. This property of pressure can be used to store large amounts of gases in smaller tanks full of liquefied gas.

Example: A gas in a closed container exerts a force of 550 N on a container that has an area of 2 m^2. Find the pressure of the gas on the container.

To solve this problem, we use the formula for pressure:

$p = F / A$

p = pressure in Pascals

$F = 550$ N

$A = 2\ m^2$

$P = 550 / 2 = 225$ Pa

Answer: The gas exerts a pressure of 225 Pa on its container.

Concept Reinforcement

1. What are the units of pressure?

2. As an airplane increases in altitude, what happens to the pressure of the air in the airplane?

3. A gas in a closed container exerts a force of 330 N on a container that has an area of 3 m^2. Find the pressure of the gas on the container.

4. An industrial press has an area of 200 in^2 over which it exerts force to press steel plates. If the press is operated with a force of 40,000 pounds, what is the pressure exerted by the press in psi?

Section 1.15 – Applications

Section Objectives

- Describe an application of Chemical Engineering

Applications of Chemical Engineering

Motor oil – one of many products of oil refining

Petroleum and its Products

Many chemical engineers work in the petrochemical industry. These chemical engineers are responsible for the refining of crude oil, also known as **petroleum**, into the various end products that are used by people around the world.

Refining is a term meaning to purify a substance. When petroleum is mined out of the ground, it is a mixture of many different compounds called hydrocarbons. **Hydrocarbons** are large molecules that are made up of a chain of carbon atoms with hydrogen atoms attached to the outside of the chain. The process of refining separates these hydrocarbons based on the length of their carbon chains, so that shorter chained hydrocarbons can be isolated from longer chained hydrocarbons. The chemical bonds of hydrocarbons store a lot of energy. Many hydrocarbons are useful as fuels or energy sources, such as gasoline, natural gas, and kerosene.

As an unrefined mixture, petroleum is not very useful for anything in particular. But when petroleum is refined into its separate components, each of these components can be used for a number of different uses. Long chain hydrocarbons form substances like tar and motor oil. Medium length hydrocarbons form substances like heating oil and diesel fuel. Short chain hydrocarbons form very volatile substances like gasoline, kerosene, and propane.

Oil refining

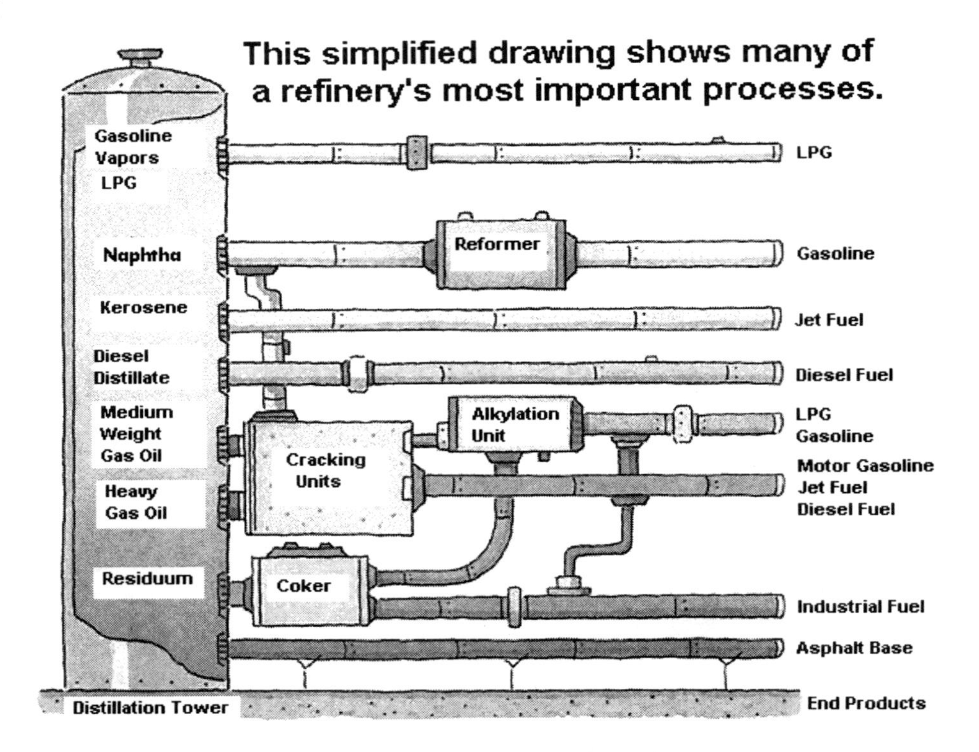

Oil refining processes and products

A refinery is a very complex structure containing many processes. Starting with crude oil, chemical engineers need to purify it and refine it down to motor oil. First, the viscous liquid crude oil is pumped into a tank. Engineers know how much crude oil is in the tank because they monitor the flow rate of the oil into the tank. As the oil arrives at the refinery, it is tested to determine its constituents so that engineers can set up the basis parameters to refine the oil.

At this point the refining process begins by heating the tank which increases the pressure of the tank. Engineers keep measuring the temperature and pressure so that they know what is happening in the tank. As the temperature increases the lightest substances based on molecular weight begin to boil and rise in the tank. These substances are the hydrocarbons with the shortest carbon chains.

The gas is then released into a large fractional distillation column, which separates the gas into its various constituents. From these early fractions, we obtain products such as methane, propane, and gasoline. As the temperature and pressure increases, heavier fractions come out of the tank and through the column. These are products such as kerosene and diesel. Additional heat and pressure drive out heavier product such as the motor oil.

All of these products are then tested using various techniques:

Chromatography and spectroscopy–to determine the purity and concentration
Physical parameters – density, boiling point
Gravimetric analysis – impurities

Petro Canada Oakville Refinery

Uses of Oil

The most obvious use of oil is for fuels. When gasoline, kerosene, diesel fuel, fuel oil and natural gas are heated in the presence of air, they will all combust. The process of combustion releases the energy stored in the bonds by the hydrocarbons. Hydrocarbons are good fuel sources because they are very energy dense – just a small amount of a hydrocarbon contains a large amount of energy in its chemical bonds compared to other fuel sources.

The products of oil refineries are used for much more than just fuels. For example, oil provides the raw material needed for the production of plastics. Chemical engineers have found ways to chemically bond hydrocarbons into even longer chains that we call plastics. One of the most important reasons for recycling plastics is to cut down on the amount of oil used to make plastics and to keep the price of plastics low. As the cost of oil rises, so will the cost of the plastic items that are used daily by people around the world.

Finally, oil provides many other chemicals besides strict hydrocarbons. They chemicals are known collectively as petrochemicals and are used in a wide variety of processes. Xylene and benzene are petrochemicals that are used as solvents. Nylon and polyester are petrochemicals that are used to make cloth for clothing. Many of the products that we commonly enjoy in our daily lives are made from petrochemicals that are produced by oil refining.

Concept Reinforcement

1. What is petroleum?
2. How is petroleum separated into its components?
3. How do we measure purity in the final products?
4. What kinds of products are made from petroleum refining?

Unit Two

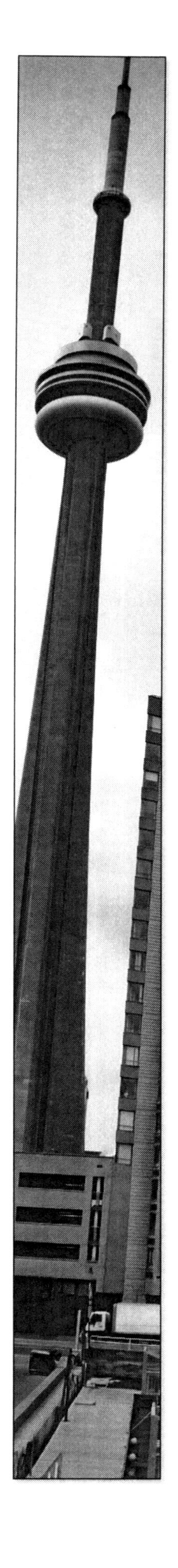

Section 2.1 – Material Balance

Section Objective

- Explain the concept of material balance

Material Balance

Material balance is one of the most basic concepts in chemical engineering. In an industrial process, engineers track many parameters such as flow rate, temperature and pressure. Another parameter that is tracked is the amount of material that is used in the process, along with the amount of final product that is produced. This is material balance. **Material balance** states that what goes in must come out.

The **Law of Conservation of Matter** states that matter cannot be created nor destroyed, but remains constant. Therefore, everything that is put into the process must be accounted for at the end. Material balance is the application of the Law of Conservation of Matter.

Why is material balance important?

Material balance is an important part of any chemical process. It helps us to measure the efficiency of chemical processes. A defined amount of starting chemicals, also known as **reactants**, is always used in a chemical process. A final product is formed by combining these reactants in the right proportions with parameters, such as heat and pressure. This product must meet specific compositional criteria to be correct. In other words, the product must be the correct product and must be pure enough to be usable.

Also, by looking at the final product, we are able to determine how much of the starting material remains after the process is complete. If we can increase the efficiency of our process, we can increase the amount of final product formed and reduce the amount of starting material that is left over.

Possible losses of starting chemicals:

Leak in the system caused a loss

Inefficient process cause waste in the form of pollutants

Inefficient process causes final product to not meet the right specifications

Perfect material balance is achieved when all of the starting material is used to create the final product. For example, if you want to make table salt, you need to combine sodium (Na) and chlorine (Cl). If you do this in the proper proportions, in this case, one to one, you will not have any leftover Na or Cl. However, if you have 25% more Na than you need, you will have Na left over after all of the Cl has been used up.

Calculating Material Balance

The formula for determining material balance is:

Input – Output = Waste

In an ideal situation, the mass of the reactants input into a system is equal to the mass of the product that is output from the system, so the mass of waste is 0. However, this is rarely possible in the real world. Even if chemicals are mixed in the exact ratio, some reactants will not interact during the reaction and will be lost as waste product.

Example: Let's say that a chemical plant is making a product from 4 different reactants. If 100 kg of reactant 1, 150 kg of reactant 2, 50 kg of reactant 3 and 100 kg of reactant 4 are used and 385 kg of product is produced, how much material is lost as waste product?

To find the answer, we need to add the masses of the reactants, and then subtract the mass of the product.

Mass of reactants (input) = 100 kg + 150 kg + 50 kg + 100 kg = 400 kg

Mass of product (output) = 385 kg

Difference between input and output = 400 kg – 385 kg = 15 kg

Answer: 15 kg of material is lost as waste somewhere between the input and the output. This waste could be accumulated within the plant, it could have been removed to a waste processor, or it could have left the plant as pollution in the air or water.

Concept Reinforcement

1. What law states that matter can neither be created nor destroyed?

2. Other than leaks in the system and products not meeting specifications, what is another way that chemicals are lost in a process?

3. Explain what happened if you have starting material remaining after the chemical reaction is complete.

4. A chemical plant is making a product from 4 different reactants. If 200 kg of reactant 1, 75 kg of reactant 2, 125 kg of reactant 3 and 300 kg of reactant 4 are used and 680 kg of product is produced, how much material is lost as waste product?

Section 2.2 – Open & Closed Systems

Section Objective

- Describe open and closed systems

What is an open system?

A **system** is defined as a group of interacting units that are related in some common purpose. Systems can then be defined as open or closed.

An **open system** is a system that interacts with the environment or other systems. In open systems, there is a continuous flow of energy or material into and out of the system process. Almost all of the systems that chemical engineers work with are open systems because they have inputs and outputs of energy and materials. The goal of most engineering projects is to produce a product, so the system must be open to allow the product to be delivered.

An example of an open system is the human body. Material and energy enter the system while energy and material exit the system in a different form. The human body gets energy from the food it consumes, converting it to energy to do things like walking and running. It also uses energy for basic life processes, like digestion, breathing, and blood circulation. Chemical plants operate like the human body – they take in energy and materials, use the energy to perform work, and produce products.

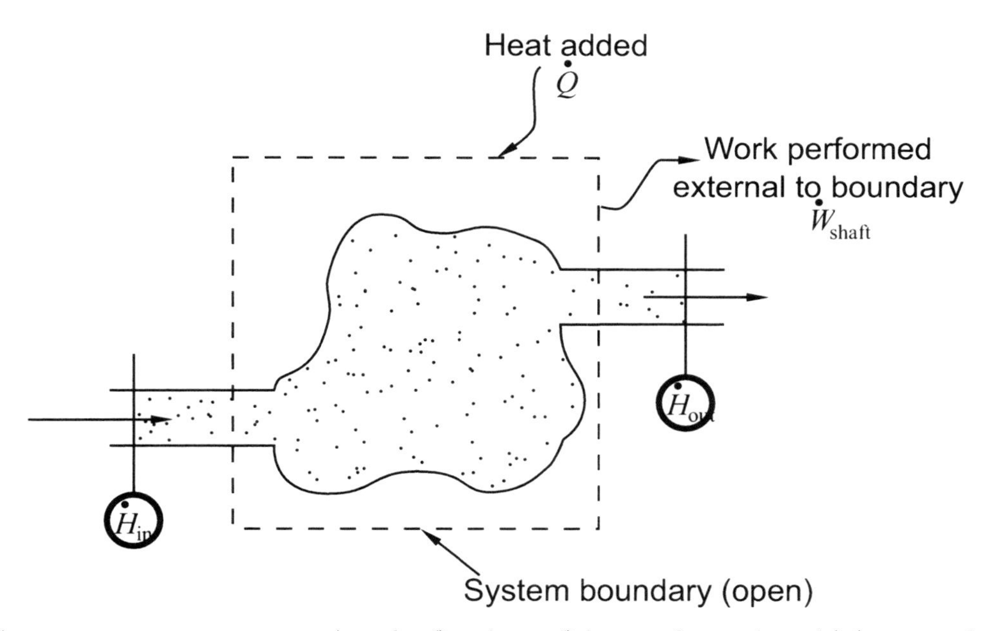

Diagram of an open system representing the first law of thermodynamic, which states that energy is never lost nor gained, but remains constant. In an open system, energy enters and leaves the system.

What is a closed system?

A **closed system** is a system that is isolated from the environment and other systems. It is also characterized by having feedbacks. In an industrial setting, it is almost impossible to have a completely closed system, since there must be materials and/or reactants added from the outside.

An example of a closed system is in the pharmaceutical manufacturing industry. In this type of system, the chemical components are added to a reactor and it is sealed. The contents are then placed under pressure at a certain temperature and allowed to react for the required amount of time. Once the reaction is complete, the system is opened to remove the finished product.

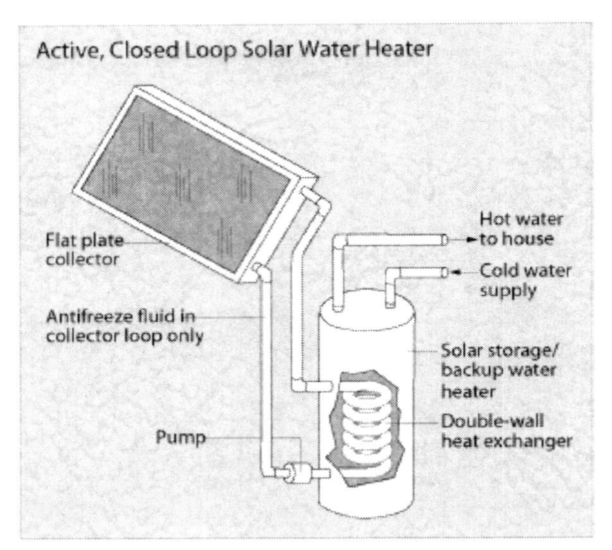

Diagram of an active closed-loop solar hot water system

Image courtesy of the US Department of Energy

Concept Reinforcement

1. Why is the human body considered to be an open system?
2. Give an example of a closed system.
3. Why is it rare to have a closed system in an industrial setting?

Section 2.3 – Steady-State and Unsteady-State Systems

Section Objective

- Describe steady-state and unsteady-state systems

What are steady-state systems?

Steady-state systems are systems where all the variables of the system do not change but instead remain constant over time. In this type of system, all parameters remain the same throughout the process. If you analyze the process at a certain point, and come back to that point at a later time, the results would still be the same.

Engineers working with steady-state systems use computerized sensors to monitor the variables within the system. If any of the variables changes to a value outside of its acceptable range, changes have to be made to the system or the system has to be shut down to fix whatever is taking the variable out of its steady-state.

An application of a steady-state system is the manufacturing of steel cable. In this instance, small pieces of wire are woven together into large spools of cable. The process runs continuously unchanged until the desired length of cable is reached.

Metal cable

Image courtesy of the Oak Ridge National Laboratory, US Department of Energy

What are unsteady-state systems?

Unsteady-state systems, also known as **transient systems**, are systems where variables change over time. This means that at any time in the process factors such as the concentration of the reactants or the temperature and pressure may vary.

Many steady-state systems will temporarily be unsteady-state systems when they are first starting up or shutting down. As a system comes online, it takes some time to reach its optimal operating conditions. During this time the product produced by the system may not be as consistent and more waste than usual may be produced as a side-effect of variables being out of their normal operating ranges. For this reason, steady-state systems are often left running as long as possible to minimize the waste associated with shutting down and starting up again.

An application of unsteady-state system is the making of wine. Grape juice, water and yeast are combined in the vessel. As the wine ferments over time, the amount of grape juice decreases, while the amount of alcohol and carbon dioxide increase.

Chateau de Lussac's Oak barrel cellar for storing wine

Concept Reinforcement

1. What is an example of a steady-state system?
2. What is an example of an unsteady-state system?
3. What is the primary difference between a steady-state and an unsteady-state system?
4. Why is it important to keep all of the variables the same in a steady-state system?

Section 2.4 – Multiple Component Systems

Section Objective

- Describe multiple component systems

What are multiple component systems?

In most chemical plants, there are usually multiple chemicals being used as reactants, multiple products being produced, and multiple processes needed to produce the final products. Systems that are made up of more than one process are known as **multiple component systems**. Multiple component systems are much more complicated than simple single component systems because there are many more factors that need to be monitored and accounted for.

In these circumstances, there are a variety of processes running at different times with chemicals being added in different stages. For instance raw chemicals are blended in a reactor. The reactor is heated and a chemical reaction occurs. The reactant is blended with another chemical producing a secondary product along with a gas as a by-product. The gas is pulled off, purified and sold. The secondary product is distilled, producing the final product along with additional by-products.

Due to the complexity of most chemical production processes, it is much more common to find multiple component systems within a chemical plant than it is to find single component systems. A large part of the job of the chemical engineer is to work with these complex systems and keep track of all of the factors involved in these systems.

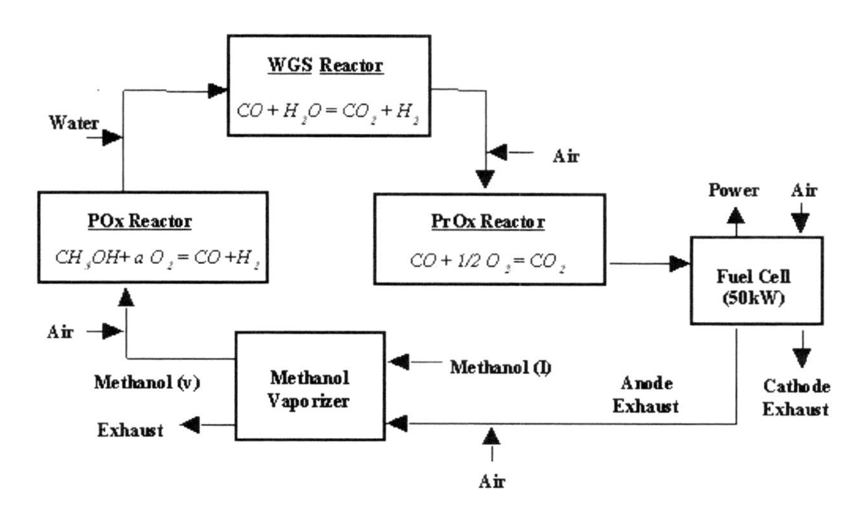

Diagram of a multi-component system

Image Courtesy of Pacific National Laboratories, US Department of Energy

Example of a multi component system

An example of a multiple component system is a fertilizer plant. Phosphate ore is mined along with a mixture of sand and clay. These components are mixed with water to form a slurry. Using physical means, the ore is separated from the sand and clay. In this scenario, liquid sulfur is burned to produce sulfuric acid. This process generates significant heat which is captured and converted to steam. The steam is then used to power the plant. The sulfuric acid that is produced is reacted with the phosphate ore to produce phosphoric acid. A by-product of this reaction is gypsum, which is stockpiled for use as plaster and drywall. The bulk of the phosphoric acid is then reacted with ammonia to produce fertilizer. Other products are commercial grade phosphoric acid for use in other chemical processes and animal feed for livestock.

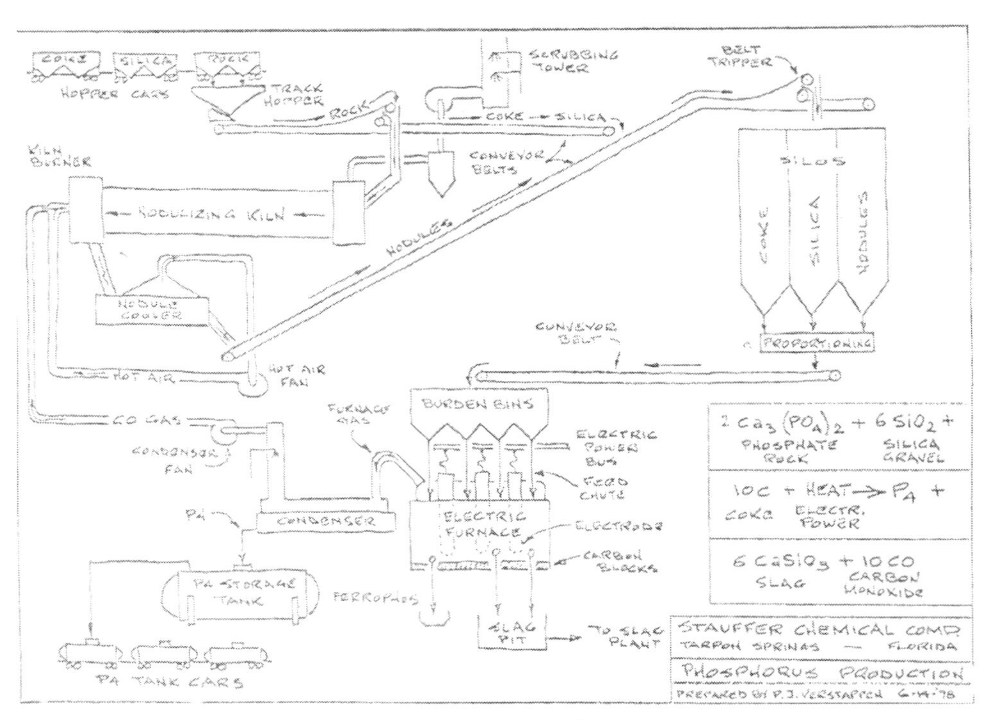

Phosphorous production flow diagram

Image courtesy of the Centers for Disease Control, US Government

What are the possible components that can be used?

In the above example, the following processes were used:

Screening to physically separate the mixture

Combustion to produce sulfuric acid

Steam production to generate power

Chemical blending

Chemical purification

As you can see, there are multiple processes to produce a final product such as fertilizer. It is important to monitor each part of the process to ensure that the final product is what it is supposed to be.

Concept Reinforcement

1. List four processes used in a multi-component system.
2. Why are multiple component systems more common than single component systems
3. How can a multiple component system be able to produce more than one output or product?
4. Why is it important to monitor each step in a multiple component system?

Section 2.5 – Chemical Reactions and Material Balance

Section Objective

- Explain how to account for chemical reactions in material balances

Chemical Reaction

A **chemical reaction** is a process where two or more chemicals are combined to create a new substance. The chemicals that are combined in the reaction are called **reactants**. The substances made as a result of the chemical reaction is called the **products** of the reaction. The products can be physically and chemically different from the properties of the reactants.

An example of a simple chemical reaction is the combination of two gases, hydrogen and oxygen, to create the liquid compound water. Hydrogen gas combusts when it is ignited in the presence of oxygen gas to form water. The product is chemically and physically very different from the reactants that are used to make the product.

$$2H_2 + O_2 \rightarrow 2H_2O$$

The Water Equation

Balancing Chemical Reactions

When working with chemical reactions, chemists and chemical engineers must always be sure that the equations that they are working with are balanced. A **balanced chemical equation is** an equation that represents a chemical reaction in which all of the atoms that are present in the reactants are present in the same amount in the product. In other words, a balanced chemical equation is an equation that takes into account the Law of Conservation of Matter – matter cannot be created or destroyed. So whatever matter goes into a chemical reaction as reactants must also come out of the reaction as a product.

In the water example, we know that hydrogen gas (H_2) and oxygen gas (O_2) combine to form water (H_2O).

$$H_2 + O_2 \rightarrow H_2O$$

Notice that there are 2 hydrogen atoms and 2 oxygen atoms on the reactants side of the equation (the left side of the arrow) and 2 hydrogen atoms and 1 oxygen atom on the products side of the equation (the right side of the arrow). This is not a balanced equation because an oxygen atom is missing. To balance the equation, we must change the ratio of hydrogen and oxygen being reacted so that all of the atoms are balanced. In this case, we will double the number of hydrogen gas molecules in the reaction. This will give us a total of 4 hydrogen and 2 oxygen atoms in the reactants. This gives us enough hydrogen and oxygen to make 2 water molecules. The equation is now written as:

$$2H_2 + O_2 \rightarrow 2H_2O$$

This is now a balanced equation. The numbers written in front of the hydrogen and water molecules are known as **coefficients**. The coefficients tell us that there are 2 hydrogen molecules and 2 water molecules for every 1 oxygen molecule.

Example: An example of a chemical reaction is the reaction between Zinc metal (Zn) and hydrochloric acid (HCl). This reaction is written as:

$$HCl + Zn \rightarrow ZnCl + H_2$$

This equation is not balanced because there is only 1 hydrogen atom in the reactants and 2 hydrogen atoms in the products. To balance the equation, we will double the number of HCl, Zn, and ZnCl molecules being used in the reaction by writing the coefficient 2 in front of the molecules.

$$2HCl + 2Zn \rightarrow 2ZnCl + H_2$$

This is now a balanced equation. There are now equal numbers of hydrogen, chlorine, and zinc on both sides of the equation. This equation tells us that for every mole of hydrogen, chlorine, and zinc that we combine, we will produce 2 moles of zinc chloride and 1 mole of hydrogen gas.

Accounting for Chemical Reactions in Material Balances

In chemical engineering, we use chemical reactions to produce the final products that we are interested in. So how do we keep track of all the chemical changes that happen through the reactions? Through chemistry, we know what the reaction will be when we combine various chemicals. As we showed above, reacting hydrogen and oxygen make water. As we see in the equation, we need to have 2 molecules of hydrogen (H_2) to react with one molecule of oxygen (O_2). This reaction creates 2 molecules of water (H_2O).

We know how much starting products we have. We have an analysis of the various starting products so we know the amounts of these products. For the final products, we can analyze them to determine the quantities of the various constituents. We are also able to measure the concentration of various by-products. Adding all these different sections should give us the proper material balance.

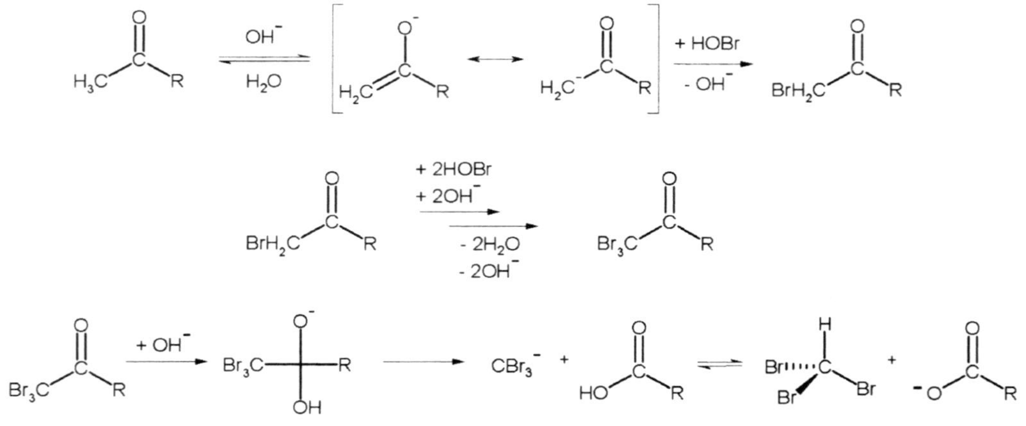

A schematic of a haloform chemical reaction

Concept Reinforcement

1. What are the initial chemicals used in a chemical reaction called?
2. Material balance in a chemical reaction comes from starting products, finals products, and what other type of product?
3. Balance these chemical equations:

   a. $Al + O_2 \rightarrow Al_2O_3$

   b. $NaOH + H_2SO_4 \rightarrow Na_2SO_4 + H_2O$

   c. $NH_3 + O_2 \rightarrow N_2 + H_2O$

Section 2.6 – Material Balances in Batch and Semi-Batch Processes

Section Objective

- Explain how to calculate material balances for batch and semi-batch processes

Batch and Semi-batch processes

There are three types of processes used in chemical reactions. These processes are batch, continuous and semi-batch.

Batch – A batch process is used when the reactants are placed in a vessel of some type and the reaction occurs. The resulting final product is removed from the vessel and more reactant is added. This process is repeated until enough of the final product is created.

An example of a batch process is baking bread. All the ingredients are mixed together. The dough rises in the pan. The bread is baked, removed from the pan. Fresh ingredients are added to the pan and the whole process starts again.

Continuous – A continuous process is where reactants are constantly added, and the final product is constantly removed. There is no stopping at all in the process.

An example of a continuous process is found in the engine of a car. Gasoline is constantly pumped into the engine where it is mixed with air for combustion. The combustion releases power. This continues constantly as long as there is gas and air.

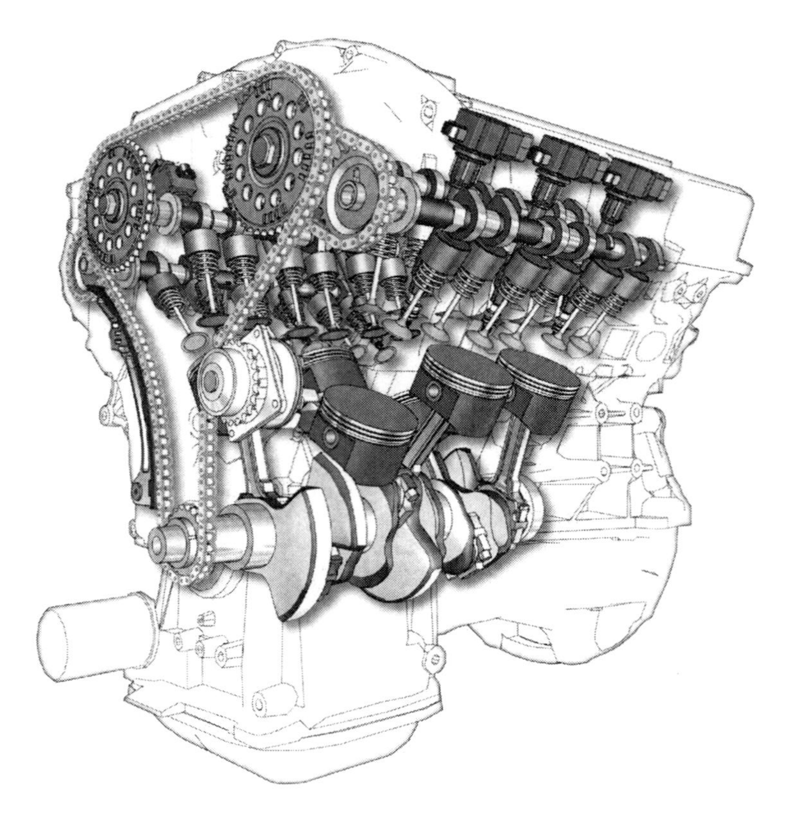

A combustion engine

Semi-batch – A semi-batch process is a hybrid of the continuous process and the batch process. There are both batch and continuous aspects to the process.

An example of a semi-batch process is fermentation. In this process, the ingredients for fermentation, say of beer, is added to the fermenter. This is a batch of beer. Through the process of fermentation, the beer generates carbon dioxide which is continuously removed from the system.

Picture of beer fermenting

Calculating material balances for batch and semi-batch processes

To calculate material balances we need to determine the quantities of raw materials required and the by-products and final products obtained.

For batch processes this is an easier task since we have a sealed system and we know what the original reactants and the final product are. By calculating how much of the final product theoretically should be made versus the actual amount of the final product produced, we get a number that represents the by-products or un-reacted reactants.

The analysis is the same for semi-batch processes, but in addition we would have to take into account any additional reactants added or by products that were removed. The formula for finding the material balance for batch and semi-batch processes is:

Input – Output = Waste

Example 1: A batch of 350 kg of pesticide is made in a chemical factory. There are 4 reactants used to make the pesticide: 200 kg of reactant 1, 50 kg of reactant 2, 75 kg of reactant 3, and 50 kg of reactant 4. How much mass was lost as waste?

To find the answer, we need to add the masses of the reactants, and then subtract the mass of the product.

Mass of reactants (input) = 200 kg + 50 kg + 75 kg + 50 kg = 375 kg

Mass of product (output) = 350 kg

Difference between input and output = 375 kg – 350 kg = 25 kg

Answer: 25 kg of material is lost as waste in each batch somewhere between the input and the output.

Example 2: A 435 kg of beer is made in a brewery using a semi-batch process. There are 4 reactants used to make the beer: 300 kg of water, 120 kg of grain, 20 kg of yeast, and 35 kg of hops. Carbon dioxide gas is released during the process of fermentation. How much mass was lost as waste?

To find the answer, we need to add the masses of the reactants, and then subtract the mass of the product.

Mass of reactants (input) = 300 kg + 120 kg + 20 kg + 35 kg = 475 kg

Mass of product (output) = 435 kg

Difference between input and output = 475 kg – 435 kg = 40 kg

Answer: 40 kg of material is lost as waste in each batch somewhere between the input and the output. Some of this mass is released as carbon dioxide gas.

Concept Reinforcement

1. What is an example of a batch process?
2. What is an example of a continuous process?
3. What is an example of a semi-batch process?
4. A batch of 650 kg of a plastic is made in a chemical factory. If there are 3 reactants used to make the plastic and 255 kg of the first reactant, 240 kg of the second reactant, and 185 kg of the third reactant were used in the reaction, how much mass was lost as waste?

Section 2.7 – Material Balance Problems

Section Objective

- Solve material balance problems

Material Balance Problems

Material balances are an important step in designing new processes or checking on existing processes. Material balance equations are a form of accounting using the law of conservation of mass that is used frequently in chemical engineering. The basic equation used in material balance is:

Input – Output = Waste

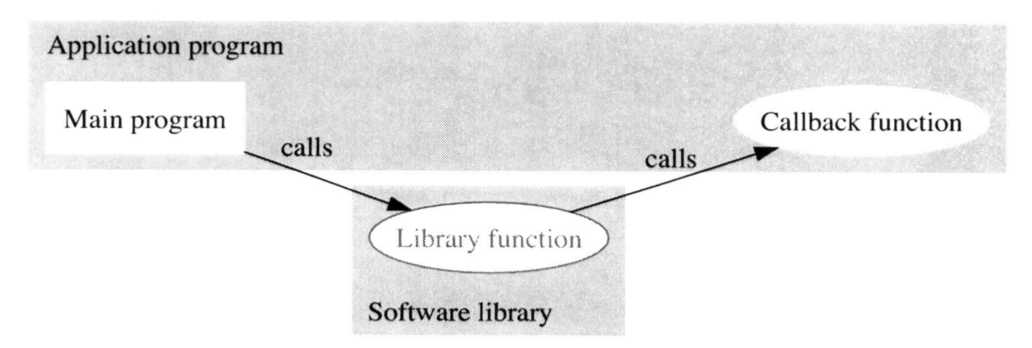

This shows a basic flow chart. A more detailed flow chart will show the specific functions that occur within the main program, the library function, and the callback function.

To solve a complex material balance problem, it is important to use a process flow chart. A **flow chart** is a graphical representation of a process that shows how materials flow from one part of a process to another. All of the steps involved in the process are included in the flow chart, with arrows showing the direction of flow of materials between each step of the process.

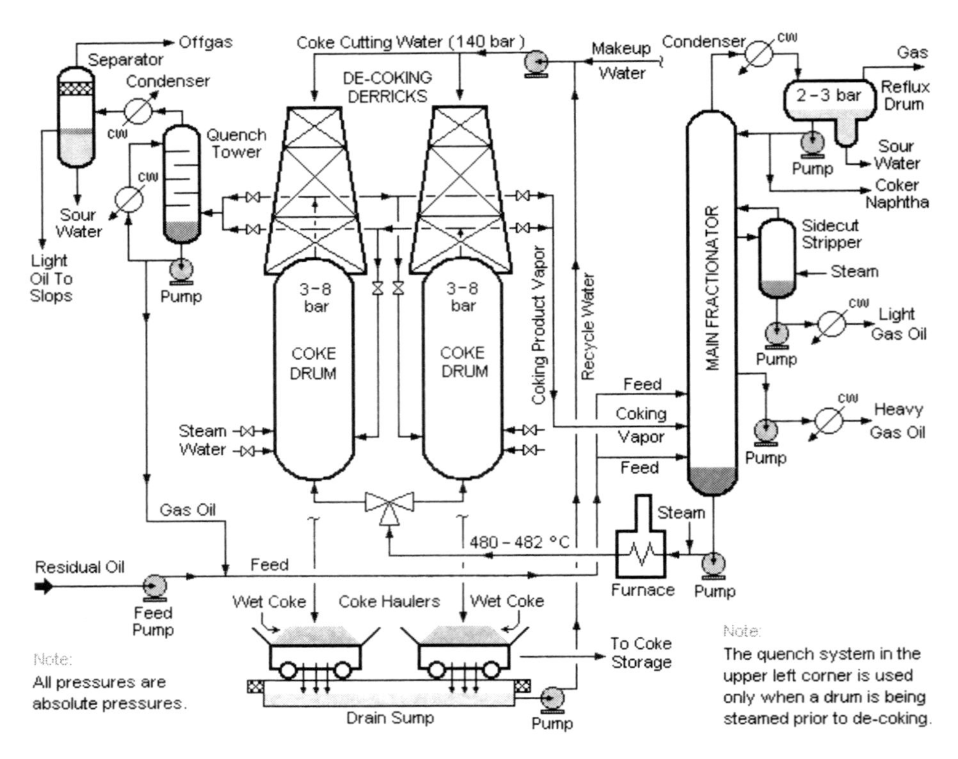

This is a detailed flow chart of a chemical engineering process. Notice that pressures, temperatures, flow direction, inputs/outputs, and systems within the process are all detailed in this flow chart.

Solving Material Balance Problems

There are four steps necessary in performing a mass balance equation. These steps are:

1. Draw and label a process flow chart.

2. Select a basis for the calculation. This means that what units are you going to use in the calculation? How are the flow rates calculated? How is the mass presented?

3. Write a material balance equation.

4. Solve the equation for the unknown quantities.

Example: An acid plant needs to make 1000 kg of 33.5% sulfuric acid for an acid battery manufacturer from diluting concentrated sulfuric acid (97%) with distilled water. The sulfuric acid flows into the plant at a rate of 86.34 kg/hr and water flows into the plant at a rate of 163.66 kg/hr. Calculate the following:

1. The amount of sulfuric acid we need to start with to make 33.5% sulfuric acid.

2. The amount of water necessary to make 1000 kg of 33.5% sulfuric acid.

3. How long will it take to manufacture 1000 kg of sulfuric acid at the stated flow rates?

Answer: First we need to identify the variables that we will use in this problem, make a flow chart, and choose a basis.

X_1 = kg of sulfuric acid (H_2SO_4) input

X_2 = kg of water (H_2O) in

t = Time (in hours)

Flow chart:

X_1 kg of 97% H_2SO_4 @ 86.34 kg/hr + X_2 kg of H_2O @ 163.66 kg/hr → 1000 kg 33.5% H_2SO_4 @ 250 kg/hr

Basis: We will use kilograms (kg) to measure amounts of reactants and products and we will use kilograms per hour (kg/hr) to measure flow rate in this problem. To make calculating percentages easier, we will write percentages using the unit kg/kg, which is the same as using the % sign.

What we have to do now is balance the amount of sulfuric acid. Since matter is always conserved, the amount of sulfuric acid that goes in is equal to the amount of sulfuric acid coming out. Knowing this we have to solve for X_1.

Kg of H_2SO_4 in · Concentration of H_2SO_4 in = Kg of H_2SO_4 out · Concentration H_2SO_4 out

$X_1 \cdot 0.97$ kg/kg = 1,000 kg · 0.335 kg/kg

X_1 = 345.36 kg H_2SO_4

Now that we have the amount of sulfuric acid from the start, we can now calculate the amount of water used to dilute by solving for X_2

345.36 kg H_2SO_4 + X_2 = 1000 kg diluted H_2SO_4

X_2 = 654.64 kg H_2O

Now to determine the time that it will take you need to look at the initial flow rates.

Sulfuric Acid (86.34 kg/hr) + Water (163.66 kg/hr) = 250 kg (33.5% Sulfuric acid/hr)

250 kg/hr · t = 1,000 kg

t = 4 hours to produce 1,000 kg

Concept Reinforcement

1. What are the four steps in solving a material balance problem?
2. Using the acid plant in the example above, perform a material balance calculation to find the amount of sulfuric acid and water used as inputs. Assume that we are using the same starting concentration, but the final concentration is 10% sulfuric acid and the final volume is 2000 kg.
3. A chemical plant needs to make 500 kg of 5% sodium hypochlorite (NaClO) to be sold as household bleach by diluting concentrated sodium hypochlorite (30%) with distilled water. Sodium hypochlorite flows into the plant at 100 kg/hr and distilled water flows in at 600 kg/hr. Solve to find the following:
    a. The amount of concentrated sodium hypochlorite we need to start with to make 5% sodium hypochlorite.
    b. The amount of water necessary to make 500kg of 5% sodium hypochlorite.
    c. How many hours will it take to manufacture 600 kg of sodium hypochlorite at the stated flow rates?

Section 2.8 – Stoichiometry

Section Objective

- Define stoichiometry and explain its application in chemical engineering

What is stoichiometry?

Stoichiometry is a way of calculating the quantities of the reactants and products in a chemical reaction. It is a type of accounting where the chemical engineer is able to keep track of all the various chemical components of the reaction. This calculation can be done by weight, number of molecules or percentage of compounds.

Stoichiometry is a way of balancing chemical reactions. We know through the Law of Conservation of Matter that matter cannot be created or destroyed. We also know that elements are pure substances that cannot be changed chemically into other elements. Therefore, the number of elements in the starting reaction will be equal to the number of elements in the final product.

How do you balance equations?

Balancing equations is a simple method of ensuring that you have the same number of elements on both sides of the reaction. If we look at the simple reaction that produces ammonia, we take two different gases, nitrogen and hydrogen. These combine to make ammonia. In this reaction you have nitrogen (N_2) and hydrogen (H_2), these need to combine to form ammonia (NH_3).

$$N_2 + H_2 \rightarrow NH_3$$

Equation of unbalanced reactants

As we see, if we have one molecule of nitrogen (N_2), and one molecule of hydrogen (H_2) we do not have enough hydrogen to make ammonia (NH_3). In addition, we are left with an extra nitrogen. To make this reaction work, we need to balance the number of elements. We need to find the right ration of hydrogen to nitrogen. Since we have two nitrogen atoms (N_2), we will need a multiple of three hydrogen atoms to make ammonia (NH_3). The easiest number would be six hydrogen atoms, or 3 hydrogen (H_2) molecule. This gives us the following equation where one nitrogen (N_2) molecule combines with three hydrogen (H_2) molecules to form two ammonia (NH_3) molecules.

$$N_2 + 3H_2 \rightarrow 3NH_3$$

Balanced ammonia equation

Applications in Chemical Engineering

As illustrated in the example above, as a chemical engineer, we need to have a complete understanding of how to balance an equation. If we were operating as ammonia plant, we would need to know at what ratios we combine nitrogen and hydrogen so we are able to actually produce ammonia. If the ratios are off, we do not produce ammonia either very efficiently, or at all, and would be wasting our starting products. Stoichiometry is used when developing flow charts for chemical plants and is used as a basis for material balance equations.

These techniques apply in any chemical manufacturing. As chemical engineers, we need to be able to efficiently produce a variety of chemicals while limiting the waste of our starting materials. We also want to produce a pure product that is free from contamination by excess reactants that are present in the end product. The cost of removing impurities from chemicals is very high, so it is important to have the correct mix of chemicals from the beginning.

Concept Reinforcement

1. Balance the following equation for the production of iron chloride using stoichiometry:
   $Fe + HCl \rightarrow FeCl_2 + H_2$
2. Balance the following equation for the production of aluminum salts using stoichiometry:
   $Al + HCl \rightarrow AlCl_3 + H_2$
3. Balance the following equation for the production of silver nitrate using stoichiometry:
   $Ag + HNO_3 \rightarrow AgNO_3 + H_2O + NO$
4. What is the purpose of stoichiometry?

Section 2.9 – Material Balances for Processes

Section Objective

- Explain material balance for processes involving reactions (species, element, combustion)

How to Perform Material Balances for Processes

Material balance is the process of applying the Law of Conservation of Matter to industrial applications. Performing material balance calculations for a variety of processes, including species, elements, and combustion, is basically the same for each process.

Material Balance Equations for Species

A **chemical species** is a general term for a molecule or atom involved in a reaction. In calculating a material balance equation for species, you need to account for all elements, regardless where they come from. In the following example we are combing two gases, carbon monoxide (CO) and hydrogen (H_2). The end products for this reaction are octane (C_8H_{18}) which is a liquid, and water (H_2O).

$$CO + H_2 \rightarrow C_8H_{18} + H_2O$$

As we look at the unbalanced equation, we see that in the end product one molecule of octane will need eight carbon atoms. To balance this, we need to start with eight molecules of carbon monoxide (CO). These eight carbons will also give us eight oxygen atoms which will have to be incorporated into the water molecules

$$8\ CO + H_2 \rightarrow 1C_8H_{18} + 8H_2O$$

This leaves us with how many molecules of hydrogen (H_2) we need. Looking at the end products we see that we have 8 molecules of hydrogen (H_2) in the water, and eighteen atoms or 9 molecules of hydrogen (H_2) in the octane. Combing these two we get 17 hydrogen molecules to balance the equation.

$$8\ CO + 17H_2 \rightarrow 1C_8H_{18} + 8H_2O$$

Material Balance Equations for Elements

Elements are pure substances, and are a specific type of species involved in chemical reactions. As with the example above, we need to balance both sides of the equation for elemental reactions. In this we react magnesium (Mg) with phosphorous (P_4) to form magnesium phosphide, an industrial product.

$$Mg + P_4 = Mg_3P_2$$

As we see in the final product, we have one molecule of phosphorous (P_4) and 3 molecules of magnesium (Mg). In this circumstance, we can see that we have 3 magnesium atoms in the final product, but only two phosphorous atoms. To balance this equation, we can assume that we will have two magnesium phosphide molecules. That will give us 6 magnesium atoms and one phosphorous (P_4) molecule. The balancing of this equation is therefore six magnesium (Mg) molecules and one phosphorous (P_4) molecule.

$$6Mg + 1P_4 \rightarrow 2Mg_3P_2$$

Material Balance equations for Combustion

Combustion is a special type of reaction where a fuel is burned in the presence of an oxidizing agent. The most common oxidizing agent used in combustion reactions is oxygen. As with the example above, we need to balance both sides of the equation for combustion processes. In this instance, we have to take into account oxygen (O_2), which is part of the combustion process. In this example, we will look at the combustion of a simple organic compound called methane (C_4).

$$CH_4 + O_2 \rightarrow CO_2 + H_2O$$

In this example we can assume one carbon on each side. This would give us then four hydrogen atoms on the final product side.

$$1CH_4 + O_2 \rightarrow 1CO_2 + H_2O$$

To make four hydrogen atoms on the final product side, we would need to have two molecules of water.

$$1CH_4 + O_2 \rightarrow 1CO_2 + 2H_2O$$

With this addition we now have four oxygen atoms on the final product side. To balance this we would need four atoms of oxygen on the reactant side which is equal to two molecules of oxygen (O_2).

$$1CH_4 + 2O_2 \rightarrow 1CO_2 + 2H_2O$$

Review: A Sample Problem

The compound sodium chloride (NaCl) is dissolved in water (H_2O) and separated by electrolysis (a strong electric current) to form the elements chlorine gas (Cl_2) and hydrogen gas (H_2), and the compound sodium hydroxide (NaOH). Write a balanced equation for this chemical reaction and find the ratio of sodium chloride to water in the reactants. Then, determine how many moles of sodium chloride and water are needed to make 100 moles of chlorine gas.

Answer: First, we will write out an unbalanced chemical equation.

$$NaCl + H_2O \rightarrow Cl_2 + H_2 + NaOH$$

Second, we will balance the equation by writing coefficients in front of each molecule. There is 1 chlorine atom on the reactants side and 2 on the products side. There are also 2 hydrogen atoms on the reactants side and 3 on the products side. To balance these elements, we will write a coefficient of 2 in front of the NaCl and another coefficient of 2 in front of the water to double the number of moles of reactants. Finally, to balance the products side, we will write a coefficient of 2 in front of the NaOH to balance the number of Na, H and O atoms. This will balance the reaction.

$$2NaCl + 2H_2O \rightarrow 1Cl_2 + 1H_2 + 2NaOH$$

Now there are an equal number of atoms of each element on both sides of the reaction. The ratio of NaCl to H_2O in the reactants is 2:2. This tells us that for every 2 molecules of sodium chloride we will need to provide 2 molecules of water to perform the reaction.

Finally, we will solve to find the number of moles of each reactant needed to make 100 moles of chlorine gas. If need 100 moles of product, we will need to run the reaction 100 times. The amount of each reactant we need is 100 reactions · the ratio for each reactant.

100 reactions · 2 mol sodium chloride/reaction = 200 mol of NaCl

100 reactions · 2 mol water/reaction = 200 mol H_2O

Another way to write this is:

$$200NaCl + 200H_2O \rightarrow 100Cl_2 + 100H_2 + 200NaOH$$

Concept Reinforcement

1. Balance the following equation for the formation of nitric acid (HNO_3) through the bubbling of nitrogen gas (N_2) and oxygen gas (O_2) into water (H_2O) using stoichiometry: $N_2 + O_2 + H_2O \rightarrow HNO_3$
2. How many moles of nitrogen gas and oxygen gas are needed to make 50 moles of HNO_3?
3. Balance the following equation for the formation of aluminum oxide (Al_2O_3) through the through the combination of aluminum (Al) and oxygen (O_2) using stoichiometry: $Al + O_2 \rightarrow Al_2O_3$
4. How many moles of aluminum and oxygen gas are needed to make 100 moles of aluminum oxide?
5. Balance the following equation for the combustion of ethane using stoichiometry: $C_2H_6 + O_2 = CO_2 + H_2O$
6. How many moles of ethane and oxygen gas are needed to make 200 moles of carbon dioxide?

Section 2.10 – Problem Solving for Material Balance with Multiple Units

Section Objective

- Solve material balance problems involving multiple units

Solving Material Balance Problems with Multiple Units

Solving material balance problems with multiple units just requires some extra levels of detail to ensure that all units are accounted for in the problem. In these instances the engineer needs to:

1. Draw a flow chart and label all streams.
2. Select a basis for the calculation. This means that what units are you going to use in the calculation? How are the flow rates calculated? How is the mass presented?
3. Separate each system by drawing a box around it. This will allow you to calculate each system separately.
4. Write a material balance equation for each separate system.
5. Solve the equation for the unknown quantities.

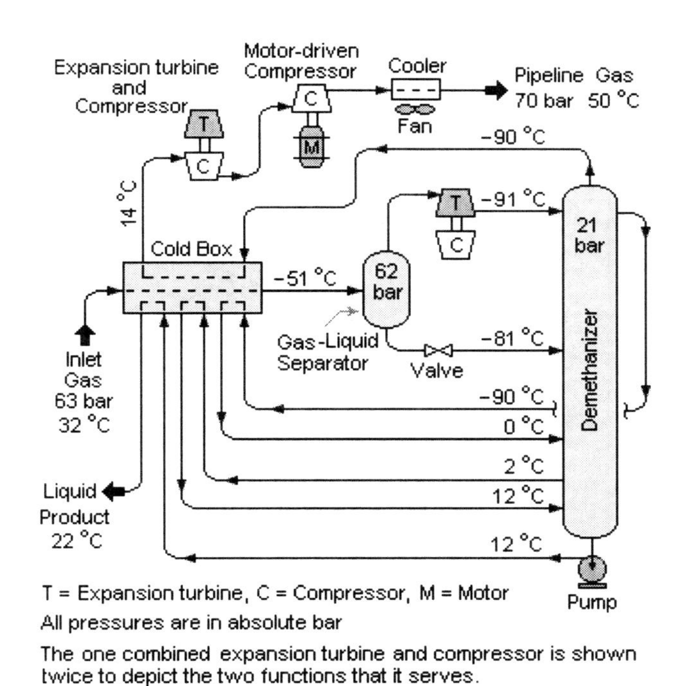

A basic flow chart

Material Balance Problems with Multiple Units

A company is manufacturing plastic widgets in a batch process. The plastic is made by mixing chemicals A and B in a reactor, which produces a liquid intermediate product and a gas by-product (System 1). The liquid intermediate is piped to a mixer where it is mixed with chemical C (System 2). The mixture is then piped to a reactor where is it heated in molds to produce the final product and another gas by-product. If 200 kg of chemical A, 100 kg of chemical B, and 145 kg of chemical C are used in the production of the plastic widgets find:

1. The mass of the intermediate piped out of System 1 if 10 kg of gas by-product is released.
2. The mass of gas by-product released from System 3 if 420 kg of product is produced.

Answer: First we need to identify the variables that we will use in this problem, make a flow chart, and choose a basis.

A = Kg of chemical A

B = Kg of chemical B

C = Kg of chemical C

G_1 = First gas by-product

G_2 = Second gas by-product

I = Intermediate

P = Product

Flow chart:

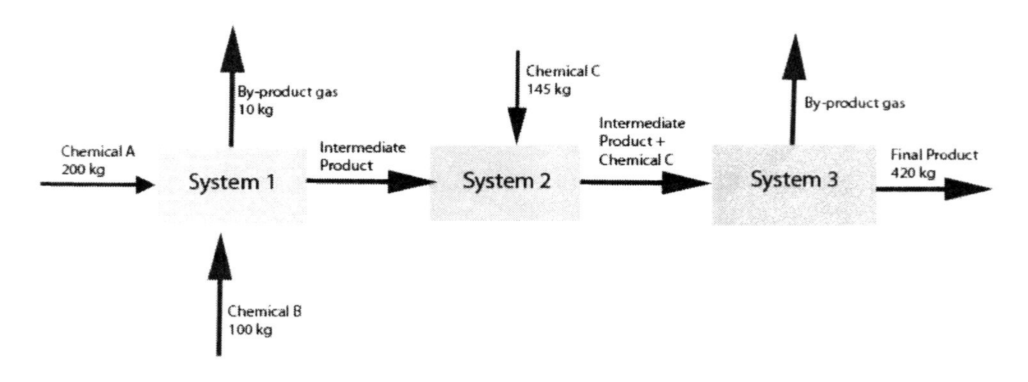

Plastic widget flow chart

Basis: We will use kilograms (kg) to measure amounts of inputs and outputs in this problem. The mass of the inputs must equal the mass of the outputs and waste, so $A + B + C = P + G_1 + G_2$

Second, we must balance the amount of intermediate that is piped out of System 1. Using the material balance equation and the amounts of each chemical, we can find the mass of the intermediate.

$A = 200$ kg

$B = 100$ kg

$G_I = 10$ kg

$A + B = I + G_I$

$200 \text{ kg} + 100 \text{ kg} = I + 10 \text{ kg}$

$300 \text{ kg} = I + 10 \text{ kg}$

$290 \text{ kg} = I$

The mass of the intermediate is 290 kg.

Third, we must calculate the mass of the second by-product gas produced in System 3. If 290 kg of intermediate is mixed with 145 kg of chemical C in System 2, then the mass of reactants entering System 3 is:

$$290 \text{ kg} + 145 \text{ kg} = 435 \text{ kg}$$

We are given the value of 420 kg for the mass of the final product. To calculate the mass of gas by-product produced, we simply subtract the mass of the final product from the mass of the intermediate entering System 3.

$$435 \text{ kg} - 420 \text{ kg} = 15 \text{ kg}$$

The mass of gas released from System 3 is 15 kg.

Concept Reinforcement

1. What are the five steps when solving a material balance problem with multiple units?

2. A company is manufacturing plastic pen caps in a batch process. The plastic is made by mixing chemicals A and B in a reactor, which produces a liquid intermediate product and a liquid by-product (System 1). The liquid intermediate is piped to a mixer where it is mixed with chemical C (System 2). The mixture is then piped to a reactor where is it heated in molds to produce the final product (System 3). If 300 kg of chemical A, 100 kg of chemical B, and 180 kg of chemical C are used in the production of the plastic widgets find:

   a. The mass of the intermediate piped out of System 1 if 40 kg of liquid by-product is released.

   b. The mass of intermediate piped out of System 2 if no mass is lost.

   c. The mass of plastic lost as waste in System 3 if 500 kg of product is produced.

3. Using the flow chart provided, solve the material balance problem for all the unknowns:

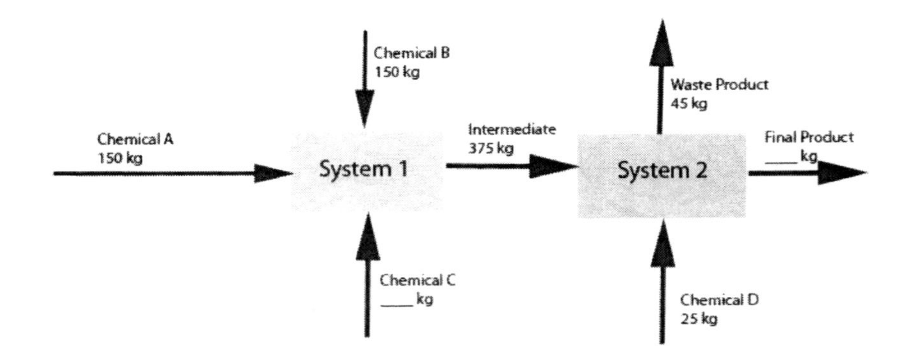

Flow chart for Concept Reinforcement question #3

Section 2.11 – Recycling Without Chemical Reaction

Section Objective

- Recycle without chemical reaction

Recycle

Recycling is the process of reusing materials in new products. We are all familiar with the concept of recycling in our homes: taking items like aluminum cans and plastic bottles that would be thrown out as garbage and reusing the items in new materials. Many of the products that we use today are capable of being recycled into new products. Also, many of the new products that we buy today are made from some recycled parts.

So, why is recycling so important? Recycling allows us to make more products from fewer materials by reducing the amount of wasted materials. Recycling is good for the environment and helps to keep the costs of materials like paper, plastic, metals, and glass low by reducing the demand for new materials.

Chemical engineers are always trying to make more products with less waste. It makes good economic sense for engineers to use recycling techniques to reduce the cost of production and reduce the amount of pollution and waste produced by a chemical plant.

The recycling symbol for PET (polyethylene terephthlate) plastic

Recycle without Chemical Reaction

In chemical engineering, you can perform processes without having a chemical reaction. These processes tend to be physical processes that can be performed anywhere throughout the entire process to make the final product. A **physical process** is a process that causes a physical change in a substance – it changes the physical, not chemical, nature of the substance. For example, mixtures have to be separated by physical means. Products have to be shaped or molded, washed, and/or packaged before leaving a plant. In these physical processes, many items can be recycled to reduce waste and keep costs low.

An example of this type of process can be found in the mining industry. In this instance, ore comes into the plant and is washed to remove any sand that might be in the ore. The slurry of sand and water is allowed to separate, and the water is recycled into the process to be used again to wash ore. The recycling of water reduces the amount of fresh water used by the plant and reduces the amount of dirty waste water that leaves the plant.

As can be seen, there is no chemical reaction going on, just a simple physical process of washing. Other processes that occur without chemical reaction are:

Distillation

Drying/Evaporation/Crystallization

Absorption

Extraction

Adsorption

Netted solution pond next to cyanide heap leaching of gold ore near Elko, Nevada

Image courtesy of the US Government

Concept Reinforcement

1. What type of process occurs during recycling without a chemical reaction?
2. Give four examples of types of physical processes.
3. How is recycling without chemical reaction accomplished in the mining industry?
4. Why is recycling important to chemical engineers?

Section 2.12 – Recycle With Chemical Reaction

Section Objective

- Recycle with chemical reaction

Recycle

Recycling is the process of reusing materials in new products. While most people think of recycling as a way of reducing the amount of garbage that makes its way into landfills, recycling is also used at the industrial level to reduce waste. In chemical engineering, industrial recycling allows higher product yields and lower amounts of waste to be produced. By making more products from fewer reactants, chemical engineers can reduce production costs and increase the profitability of a chemical plant. Recycling also helps the environment by reducing waste and pollution that is made by chemical plants.

Recycle with Chemical Reaction

Most of the processes that take place in a chemical plant involve chemical reactions between reactant chemicals to produce new products. In chemical engineering there are processes where various reactants and end products are recycled back into the reaction. They are recycled back into the process to:

Help increase the efficiency of the process

Help to speed up the process

To improve separation

To minimize cost of additional equipment and supplies

No chemical process is ever 100% efficient. There is always some matter lost as waste due to the nature of atoms and molecules being so small that they sometimes fail to mix with other particles in a reaction. There are, however, many techniques that chemical engineers can use to reduce the amount of product that is wasted. Reactants that are left over from a chemical reaction can be separated from the product and piped back to the reactor to participate in the chemical reaction again.

Waste products from chemical processes can often be recycled for use in other areas. For example, hydrogen gas produced as a by-product of chemical reactions can be collected and stored for use as a fuel supply. Hydrogen gas is becoming an important resource as engineers develop hydrogen fuel cell technology.

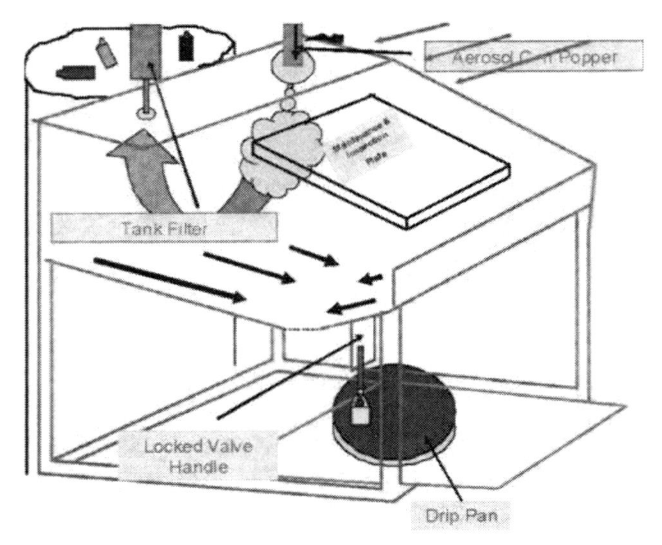

Diagram of a paint recovery system

Applications of Recycle with Chemical Reaction

An example of recycle with chemical reaction would be using a reactor to produce an organic chemical such as benzene. In this instance, the reactor produces benzene, but the production process does not produce the product fast enough. The system has a splitter after the reactor that removes some of the reactants. The benzene is separated from the other reactants with a distillation column, and then the remaining reactants are returned (recycled) to the reactor.

Fertilizer plants produce nitrate fertilizer from phosphate ore. Some of the waste products of the chemical reactions used to make the fertilizer are gypsum, phosphoric acid, and solid residues. Gypsum is a mineral that can be recycled as plaster and drywall. Phosphoric acid is a chemical that is useful in chemistry labs and in other chemical processes. The solid residues can be used in the production of concrete or gravel.

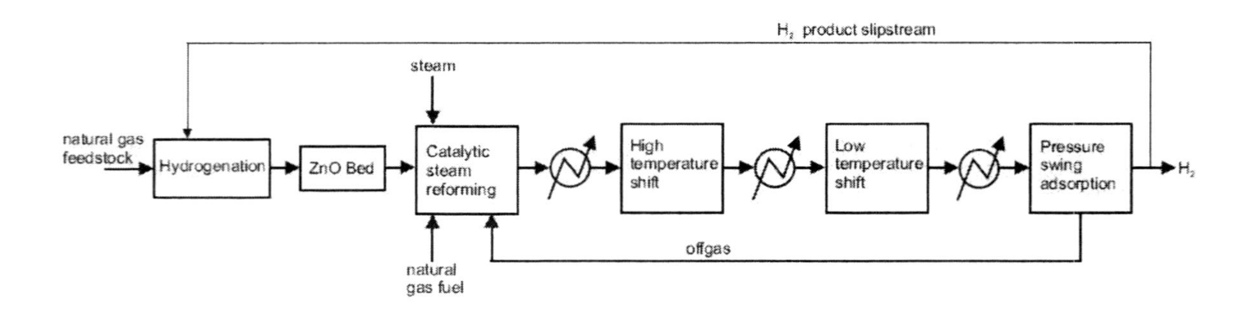

A flowchart of a chemical process that uses recycle with chemical reaction

Concept Reinforcements

1. List the four advantages to recycle with a chemical reaction.
2. What is the role of a splitter in a chemical reaction?
3. Why are chemical reactions never 100% efficient?

Section 2.13 – Bypass

Section Objective

- Describe and explain bypass

Bypass

In some chemical processes, there are systems that need to be skipped from time to time. In these cases, chemical engineers need to be able to change the flow pattern of materials through the process to avoid systems that are not needed. A **bypass** is an engineering process that skips one or more subsequent processes. Usually a bypass is made up of at least 2 valves and some pipes that run around a system. When the bypass is in use, the valve opens to allow some or all of the chemicals to flow through the bypass instead of the bypassed system.

A simple way to think of a bypass is to think of a switch connected to a lamp. When the switch is turned off, the electricity flowing to the lamp bypasses the lamp, so the lamp does not turn on. When the switch is turned on, electricity is allowed to flow to the lamp and the lamp produces light.

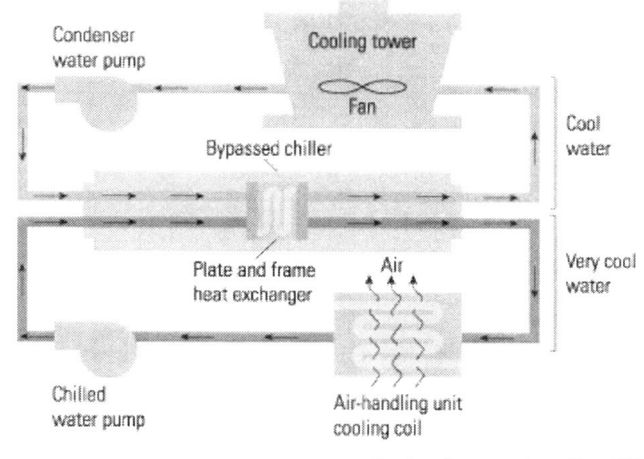

Diagram of a system bypass

Applications of Bypass

So, when are bypasses used? Bypasses can be useful in a wide variety of ways. They can be used to allow the relief of pressure in a process where pressure can build up too high. Bypasses can be used to introduce new or additional reactants later in a reaction. They can also be used to obtain precise control of the final product by adjusting how much material goes through each process.

In some cases, the same equipment can be used to make several different products. Depending on which systems are active in a process can determine which product is made. Bypass can be used in combination with recycle to pull finished product from a recycle loop. In this case the bypass works with some form of splitter to remove the product from the reactants.

In other cases, the reactants may be more or less pure, depending on their source. Pure reactants require less purification and refining, while impure reactants need to be processed more thoroughly to make a good product. Reactants that are obtained from nature, like minerals mined from the ground or agricultural products like grain tend to vary in their quality. When chemical plants depending on multiple suppliers of raw materials may find that the quality of material varies from supplier to supplier.

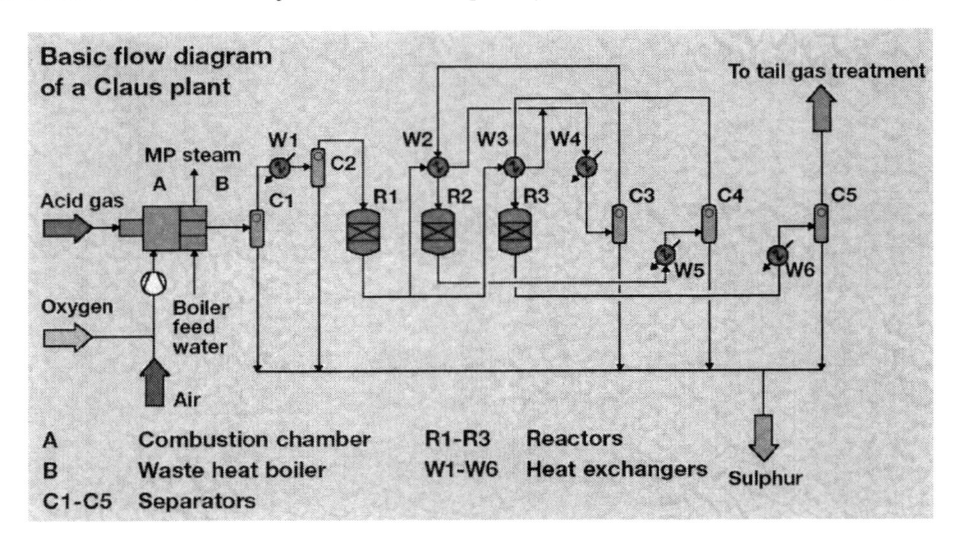

A diagram of a system for extracting sulfur from hydrogen sulfide gas. Note that sulfur can be bypassed from systems C1 through C5 once it is separated from the hydrogen sulfide gas.

Concept Reinforcement

1. What is a bypass?
2. What are three uses of bypass?
3. Why is bypass useful when working with materials that come from nature?
4. How can bypass be used in combination with recycle?

Section 2.14 – Purge and its Applications

Section Objective

- Define purge and its applications

What is Purge?

In its most general sense, the term **purge** means to remove something. In chemical engineering, purge is usually associated with removing a gas from a process stream. In some cases, a pressurized gas may be used to purge a solid or a liquid accumulation from the inside of a pipe or machinery in a chemical plant.

The material exiting a system through a purge is known as the **purge stream**. Many different types of materials may need to be purged from a chemical process. The most common type of materials to be purged from a process are gases, due to the dangerous pressures that can develop when gases are allowed to build up within a system. Gases are removed through a type of purge called a **vent**. A vent allows the gas to escape into the atmosphere or may lead to a storage tank for disposal of hazardous gases. Vents are most commonly installed in the top of equipment to allow lighter gases to flow out. Solids and liquids are removed through a type of purge called a **drain**. Drains are located in the bottom of a piece of equipment so that heavier solids and liquids can be pulled out of the system by gravity.

Purge Applications

In chemical engineering, purge has multiple applications. Some of these applications are:

1. Bleeding off from the process stream. Purging is most often done to remove excess or unwanted material that can build up in the process stream. An application of this is in an oil refinery. In this instance, unwanted gases are removed by vents throughout the distillation process. Unwanted solid particles and liquids are drained out of the bottom of the distillation column to make room for crude oil that is being added to the process.

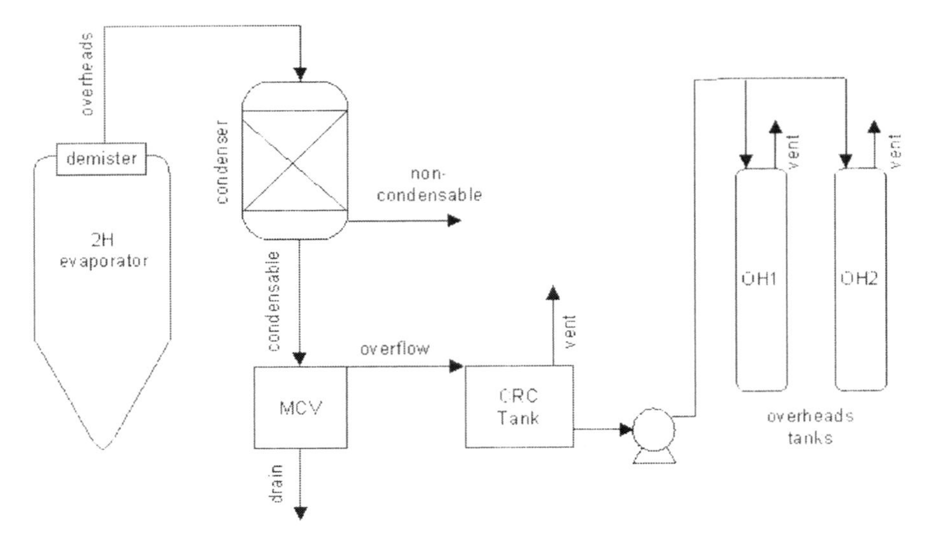

Rough schematic of the potential vent paths in the 2H-evaporator overheads system at the Savannah River site. This was part of a chemical cleaning process of the 2H-evaporator.

Image courtesy of the US Government

2. To make a tank or vessel inert. In this definition, a gas, usually an inert one is pumped into a tank or vessel forcing the gases that had previously been in the tank out. This makes the tank safe or allows a person to enter it to perform maintenance. A common application of this type of purge is in petroleum storage tanks. After a tank is empty of liquid fuel, large quantities of fuel vapor still linger inside the tank. This vapor is dangerous because it is very volatile and can cause an explosion if ignited accidentally. In addition, the levels of contaminants in the tank are too high for a person to safely enter to perform work. In this case, the tank is purged with nitrogen, which is heavier than the petroleum gas, and forces the petroleum vapor out as it fills the tank from the bottom up. Once the petroleum vapor has been purged, the nitrogen gas can be safely released into the atmosphere because nitrogen is inert and air is mostly made of nitrogen gas.

3. To keep a recycle system in balance. Recycle systems in chemical plants are very important to increase the efficiency of the chemical process, but recycle systems can also be dangerous. If too much material is being recycled and not enough material is exiting a system as product, the pressure within the system can start to rise. This is particularly dangerous in continuous systems where a continuous stream of reactants is entering the system. In this case, the recycle loop is equipped with a purge stream to remove excessive amounts of reactants. Although this material

A gasoline tanker

Concept Reinforcement

1. What are three applications of purge in chemical engineering?
2. How are liquids purged from a system?
3. Why is it important for systems that use recycle to have a purge?
4. Why is it important to purge petroleum tanks?

Section 2.15 – Applications of Chemical Engineering

Section Objective

- Describe an application for chemical engineering

Chemical Engineering Applications

Oil Refining

An application in chemical engineering is an oil refinery. An oil refinery is a complex set of systems and processes. It is an ideal place for a chemical engineer to use all the knowledge that had been gained in school.

The starting product in a refinery is crude oil. Various physical and chemical processes and reaction are performed on the crude oil to create the final products: gasoline, kerosene, diesel fuel, motor oil, asphalt, tar, petrochemicals, liquid petroleum gas (LPG), and many other products.

In the first place, the engineer knows how much crude oil is added to the system, and is also able to calculate the volume of gasoline produced. By doing a material balance analysis, the engineer is able to determine the efficiency of the process.

Aerial of an oil refinery

Image courtesy of NASA

In the refinery, many of the processes are continuous or semi-batch. Due to the high degree of automation used to refine crude oil, it is rare to have batch processes in a modern refinery. Most of the systems are open, since closed systems limit the automation.

In the refinery there are a variety of processes running at any one time. Some of these processes are:

Desalter – this process removes salt from the crude oil prior to distillation.

Fractional Distillation – this process is used to separate crude oil into its separate components based on their boiling points.

Cracking – this process is used to crack large molecules like gas oils to make smaller molecules like gasoline and kerosene.

Coking – this process takes very large, heavy molecules from residual oil and breaks it into lighter oils and fuels.

Water treatment – water used in an oil refinery is treated so that it can be recycled through the plant.

In the distillation process there are several places where the process stream is recycled back into the process to increase the efficiency of the process. There are also purge lines that help keep the temperature and the pressure of the process optimum.

As you can see there are a tremendous number of processes and systems that run in a refinery. As an engineer, you need to be able to account for all the possible variations of material balance.

Plastics Engineering

Plastics have become so common and useful in our day to day lives that most people take them for granted. Prior to the development of plastics, many common objects were made of wood and metal. While wood and metal are very good materials, they are usually more expensive than plastics and often tend not to last as long as their plastic counterparts. Plastics are incredibly cheap, have very flexible uses, are durable, and can be made into an almost unlimited number of forms. There are many different types of plastics being used today. Each type of plastic has its own advantages and disadvantages, so some are used for bottles, some for cooking equipment, and some for clothing, among other uses.

Plastic bottles

All plastics are made up of a type of molecule called a polymer. In fact, the word polymer is often used interchangeably with the word plastic, although there are some polymers that are not plastics. A **polymer** is a long chain molecule that is made up of many smaller molecules chemically bonded together. The smaller molecules that make up a polymer are called **monomers**.

Almost all plastics are made from petrochemicals, which are by-products of the oil refining process. Some of the long chain hydrocarbons that are not converted into fuels are used as the monomers to form polymer plastics. As the price of crude oil increases, the price of petrochemicals and therefore the price of plastic plastic will also increase.

Petrochemicals are shipped to chemical plants where plastics are made. The plastics are made by a chemical reaction between two or more petrochemicals along with other chemicals like acids or bases which help the plastics to form. The chemical reaction that forms the plastic is called a **polymerization reaction** because the petrochemical monomers are linking together to form a polymer.

After polymerizing, the plastic needs to be shaped into its final form. The two main processes for shaping a plastic are **injection molding**, where liquid plastic is injected into a mold to give it shape, and **extrusion**, where liquid plastic is squeezed into long sheets or tubes.

Concept Reinforcement

1. What process is rare in a modern refinery: batch, continuous or semi-batch?
2. Why are coking and cracking important processes in an oil refinery?
3. What are some advantages of making an item out of plastic?
4. How are plastics related to oil refining?
5. How are plastics made into different shapes?

Unit Three

Section 3.1 – Ideal Gases and the Ideal Gas Law

Section Objective

- Define ideal gases and explain the ideal gas law

What is an Ideal Gas?

Most of the gases that we encounter in the world around us are **ideal gases**. For a gas to be considered ideal, it must meet two assumptions. These assumptions are:

1. The space that a gas molecule occupies is significantly smaller than the space between the gas molecules.

2. The intermolecular forces between the gas molecules are negligible.

To visualize the first assumption, picture a room with a single pea in the middle of the floor. If you are asked to calculate the volume of the room it is easy to calculate this by taking the length, width and height of the room to come up with a number. Then taking into account the pea, the volume of the pea is insignificant to the volume of the room.

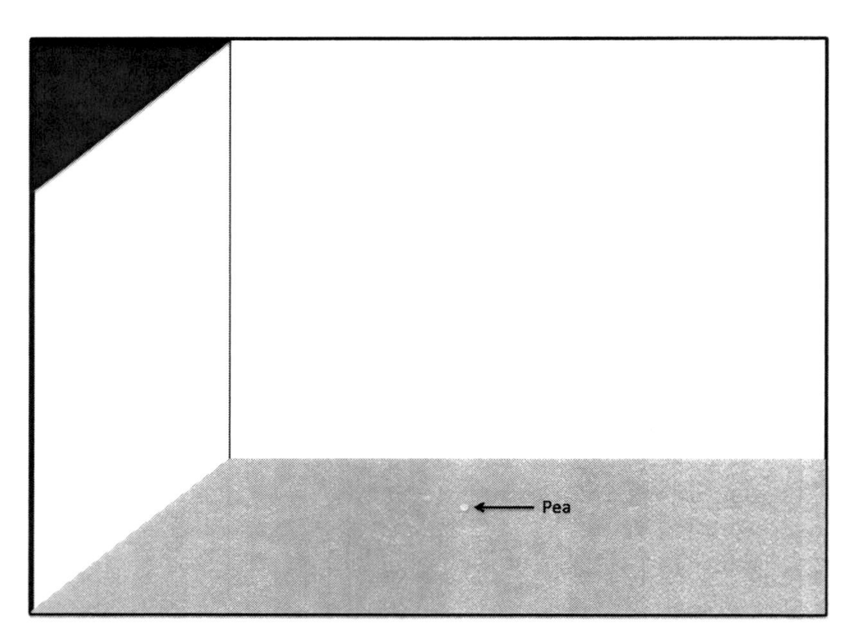

The second assumption is dependent upon the space between the molecules and the speed with which the molecules move. Although gas molecules, like other molecules, do have an attraction and repulsion for each other, due to the speed at which the molecules move, the amount of attraction or repulsion is so small that it can be ignored.

When is an ideal gas not ideal?

The two assumptions of the ideal gas law are no longer valid when the two following conditions occur:

1. Extremely high pressure – At extremely high pressure a gas is no longer ideal because the gas is compressed together, decreasing the distance between the molecules. An example of this situation is a pressurized oxygen tank. In this example the gas molecules are squeezed together in the tank under high pressure.

2. Extremely low temperature – At extreme low temperatures the gas molecules have very little energy and move very slowly. This increases the intermolecular forces between the molecules as they tend to spend more time in the same place, so they can interact with each other. Therefore, the gas is no longer considered to be ideal.

What is the Ideal Gas Law?

The **ideal gas law** is a law that links all the fundamental properties of a gas into a single equation:

$pV = nRT$

In this equation, we take into account pressure (p), volume of the gas (V), number of moles of gas (n), temperature (T) in degrees Kelvin and the ideal gas constant (R). This is a constant and never changes. This constant is usually represented as 8.3145 kPa L/mol K.

This equation is a theoretical equation since it does not take into account any intermolecular forces and assumes that all molecular collisions are perfectly elastic. But the intermolecular forces and collisions are so small that they have almost no effect on the gas.

Solving problems using the Ideal Gas Law

To solve problems using the ideal gas law, we simply plug in the values for all of the known values and solve for whichever value is unknown.

Example: What pressure would be exerted on a cylinder when 52.3 moles of oxygen are added to a 10 liter cylinder at 293 degrees Kelvin?

$pV = nRT$

$p \cdot 10 \text{ L} = 52.3 \text{ mol} \cdot 8.3145 \text{ kPa L/mol K} \cdot 293 \text{ K}$

$p = 12{,}741.056 \text{ kPa}$

Picture of a gas cylinder

Concept Reinforcement

1. What are the two assumptions that make a gas ideal?
2. What two extremes make ideal gases no longer ideal?
3. Solve the following problem using the ideal gas law: wow many moles of hydrogen can be added to a 30 liter cylinder at 1247.2 kPa and 290 K?
4. Solve the following problem using the ideal gas law: what size cylinder (volume) is needed to store 12.2 mols of nitrogen at 256.32 kPa and 372 K?

Section 3.2 – Real Gases

Section Objective

- Discuss real gases, including compressibility and equations of state

What is a real gas?

Real gases are different from ideal gases in that they do not follow the same rules as ideal gases. For a gas to be considered ideal, it must meet two assumptions. These are:

The space that a gas molecule occupies is significantly smaller than the space between the gas molecules.

The intermolecular forces between the gas molecules are negligible.

These assumptions only work for high temperatures and low pressures. When these variables change, gases are no longer considered ideal, so we call them real gases.

Compressibility

Compressibility is the ability to squeeze a gas into an area smaller than it normally resides. Think of an air tank for SCUBA Diving. In this instance you have an open tank that has air in it at standard atmospheric pressure. When you close the valve, sealing the tank, you are then able to add more air to the tank by compressing the air at high pressure.

With the air compressed, the molecules are much closer together, thus making them act in a manner that is no longer ideal. In addition, with the molecules being closer together there is more interaction between the molecules due to them being closer. As molecules of a gas are compressed, they begin to exert an outward force against the pressure that is forcing them together. This force is caused by the repulsion of the gas molecules to one another. At a certain point, it becomes more and more difficult to compress a gas any more than it is already compressed.

Real gases also begin to form liquids are certain combinations of pressures and temperatures. When the molecules move slowly and attractive forces between the molecules are great enough, the gas will begin to turn into a liquid state. Oxygen and nitrogen are examples of gases that form liquids at low temperatures and high pressures. These liquids are very cold and are used in cryogenic freezing processes. Liquid oxygen is also used as a more compact way to transport oxygen on space missions.

A compressed gas cylinder

Equation of State

An equation of state is an equation that represents the state of matter at any given conditions. It provides an equation for the relationship of various properties such as volume (*V*), temperature (*T*), pressure (*p*) and amount (*n*).

> *Boyles Law – Describes the relationship between pressure and volume. In this law, mass and temperature are held constant (pV = C). Where C is a constant.*
>
> *Charles and Gay-Lussac's Law – Describes the relationship between volume and temperature. In this law, mass and pressure are held constant (V = CT).*
>
> *Ideal Gas Law – Boyles Law and Charles and Gay-Lussac's Law are then combined to for the Ideal Gas Law (pV= nRT).*
>
> *Van der Waals Equation of State – This equation was the first of many equations to take into account molecular size and molecular interaction forces. In this equation each substance has its own constants based on it's the specific characteristics of each substance.*

Additional equations have been developed, each one perfecting the calculation a little more. But with each improvement comes increased complexity in calculating a value.

The van der Waals Equation

The **van der Waals equation** is a modification to the Ideal Gas Law that can be applied to real gases. The van der Waals equation includes variables to account for the fact that each gas act has a different way of acting at high pressures and low temperatures. The van der Waals equation is:

$(P + a/V^2)(V - b) = nRT$

where

P is pressure measured in kilopascals (kPa)

a is the correction for the molecular attraction between gas molecules

b is the correction for the volume taken up by the gas molecules

V is the volume of the gas in liters (L)

n is the amount of gas measured in moles (mol)

R is the ideal gas constant (8.3145 kPa · L/mol · K)

T is the temperature of the gas in Kelvin (K)

In order to use the van der Waals equation, the values of a and b must be known. The values of a and b are constants and are different for every gas. Here the values of a and b for several common gases:

Gas	Value of a	Value of b
Hydrogen	0.248	0.0266
Helium	0.035	0.0237
Methane	2.28	0.0428
Nitrogen	1.41	0.0391
Carbon dioxide	3.64	0.0427
Oxygen	1.38	0.0318

Example: How many moles of nitrogen would be needed to exert a pressure of 200 kPa on a cylinder when added to a 10.0 liter cylinder at 80 K?

$(P + a/V^2)(V - b) = nRT$

$P = 200$ kPa

$a = 1.41$

$b = 0.0391$

$V = 10.0$ L

$n =$ amount of gas (mol)

$R = 8.3145$ kPa · L/mol · K

$T = 200$ K

$(200 + 1.41/10.0^2)(10.0 - 0.0391) = n \cdot 8.3145 \cdot 80$

$1992.320449 = 665.16 \cdot n$

$n = 2.99524\ldots \approx 3$ mol

3 moles of nitrogen are needed to exert a pressure of 200 kPa on the cylinder.

Concept Reinforcement

1. How are real gases different from ideal gases?

2. Which two laws make up the ideal gas law?

3. How is the van der Waals equation different from the ideal gas law?

4. How many moles of oxygen would be needed to exert a pressure of 300 kPa on a cylinder when added to a 5.0 liter cylinder at 100 K?

Section 3.3 – Vapor Pressure

Section Objective

- Discuss vapor pressure, including phase diagrams and modeling vapor pressure as a function of temperature

What is vapor pressure?

Vapor pressure is the equilibrium between the gaseous and solid or liquid phase of a substance. What this means is that molecules of the substance are constantly moving between the gaseous and liquid or solid phase. As one molecule becomes gaseous, another molecule is becoming a liquid or solid. Molecules become gaseous as they are excited by energy in their surroundings and they become liquid or solid as they lose energy and return to their previous state.

The vapor pressure changes with temperature. As the temperature rises, there is more energy in the system to excite molecules to enter the gas phase. Vapor pressure increases exponentially with an increase in temperature.

An example of vapor pressure that is commonly seen in our day to day lives is a tea kettle. As the temperature of the water in the kettle rises, the increase in energy causes more water molecules to become excited and become a gas. This gas is what we call steam. As the temperature increases, the vapor pressure formed by the steam increases exponentially as much more water becomes steam. The steam begins to exert a pressure on the whistle of the tea kettle, which becomes louder as the water approaches its boiling point.

What are phase diagrams?

Phase diagrams are a graphic interpretation of what happens to a substance with respect to the different phases of matter (gas, liquid, solid). In these diagrams you are plotting temperature against pressure. What emerges is a square divided into three areas based on the three phases; gas, liquid solid.

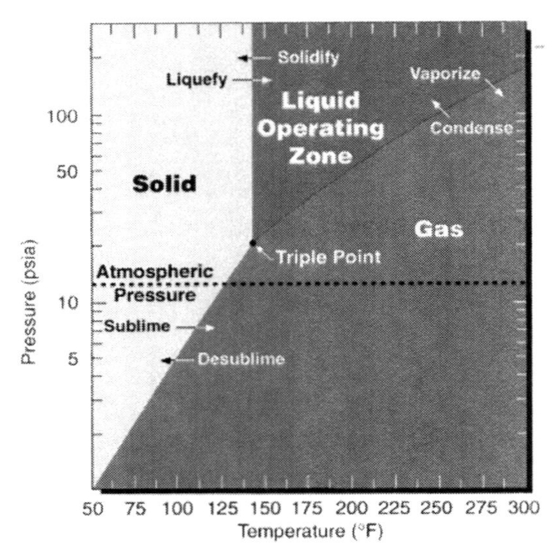

Phase diagram for uranium hexafluoride (UF_6)

Image courtesy of the Argonne National Labs

As we look at this diagram, we can see the curved lines between the phases. These lines represent the areas where equilibrium exists between the phases. Here you can have both liquid and solid, or liquid and gas, or solid and gas. The place where all three curves meet is called the triple point. In this situation, all three phases are in equilibrium. All other areas of the diagram are where only a single phase exists.

Liquid/Gas Phase

The curve between the liquid and gas is called the **vapor pressure curve**. At this point, equilibrium exists between liquid and gas. As a substance is moved from liquid to gas, it is called **vaporization**. When it moves from gas to liquid it is called **condensation**. On this curve there is a point labeled **critical point**. At this temperature, no matter how high the pressure, gas can not become liquid.

Liquid/Solid Phase

The curve between the liquid and solid phase is the equilibrium point for these two phases. A substance that moves from a liquid to solid phase is called **freezing**. When it moves from the solid to liquid phase it is called **melting**.

Solid/Gas Phase

The curve between the gas and solid phase is the equilibrium point for these two phases. A substance that moves from the solid to gas phase is said to **sublimate**. When it moves from a solid to a gas the process is called **deposition**.

Modeling Vapor Pressure

Vapor pressure can be modeled against temperature to predict vapor pressures at specific temperatures. This is useful in many processes such as distillation, adsorption and stripping. In these processes we can use various table of boiling points to help predict the temperature against pressure where the equilibrium will be reached. In doing this we can better refine our processes.

Concept Reinforcement

1. What is vapor pressure?
2. What is it called when a solid becomes a gas?
3. What is it called when a liquid becomes a gas?
4. Label all the parts of the phase diagram for water.

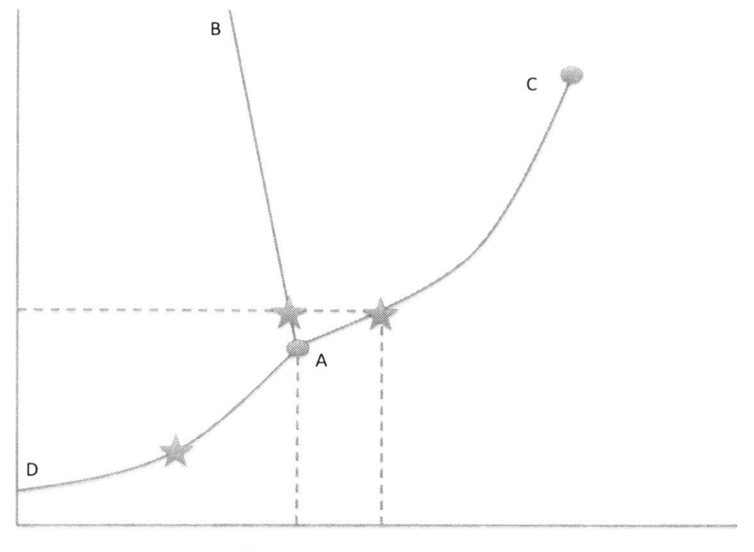

Water phase diagram

Section 3.4 – Saturation, Condensation and Vaporization

Section Objective

- Define saturation, condensation and vaporization

What is saturation?

Saturation is the condition when a mixture of both liquid and gas or liquid and a solid can exist at the same time. Saturation occurs when a one substance is dissolved in another substance to a point where you cannot get any more of the original substance to dissolve into it. A classic example of this that we all know is humidity. On a hot humid day there is a very high percentage of water vapor in the air. When humidity reaches 100% the air is saturated with water.

What is condensation?

Condensation is the point where a gas becomes a solid. When the temperature gets low enough the gas condenses, or becomes liquid. Using the example of weather that we discussed above, think again of our humid summer day. As night falls, the temperature drops, when morning arrives, the temperature has dropped low enough that the water condenses in the form of water droplets that we call dew. This temperature point is called the **dew point**, and it is the temperature at which water vapor condenses.

If the temperature drops too low, below the **freezing point**, we get frost, and instead of having condensation, we have deposition.

Dew on grass

What is vaporization?

Vaporization is the point where a liquid becomes a gas. This usually occurs at the surface of the liquid. In this case, an increase in temperature leads to an increase in vaporization. An example of vaporization is a tea kettle on the stove. The kettle is filled with water and heat is slowly applied to it. As the temperature increases, small amounts of water vapor as steam can be seen to rise off the surface. As the temperature approached the boiling point even larger amounts of steam come off until the water starts to boil and the kettle whistles due to the increase in vaporization

"A Generous Teapot" by Colonel Wedgwood

Image courtesy of the Project Gutenberg archives.

Concept Reinforcement

1. What is an example of saturation?
2. What is the point in weather where water vapor becomes liquid water?
3. Give an example of vaporization.

Section 3.5 – Partial Saturation and Humidity

Section Objective

- Explain partial saturation and humidity

What is partial saturation with respect to humidity?

Partial saturation is the condition in which a solute is present in a solution but the solution is not complete saturated. In the case of partial saturation, more solute can be added to the solution before complete saturation is achieved.

Since it is almost impossible to have complete saturation of air (100%), we need to represent it as some "partial" percentage of that 100%. The partial saturation that is most often discussed is humidity. **Humidity** is a phrase used in weather to quantify the amount of water in the atmosphere. What they are really talking about is a term called relative humidity. **Relative humidity** is the ratio of the partial pressure of water vapor to the saturated vapor pressure of water vapor. Both quantities are for the same air mass and the same temperature. In most cases the values is then multiplied by 100 to give a whole percent that people are able to understand easier.

Water droplets on a hydrophobic plastic garden table

Humidity and Engineering

Humidity effects engineering in several ways. First of all, if a facility is located in an area that has high humidity, the system may have to be designed to handle all the water vapor that is around. Will that much water vapor effect a reaction or will it damage equipment? An engineer also has to take into account any people working in that type of environment. Will it be safe for them, and how do you make it safer?

Another consideration of humidity in engineering is in the use of air from the environment in engineering processes. Air is often used for drying products and may sometime be needed for reactions, particularly combustion reactions. In these cases the humidity of the air may affect the production process.

The humidity of a volume of air can be reduced by pulling the air through a dehumidifier. There are 2 main types of dehumidifiers: ones that use a **desiccant**, or a chemical that absorbs water, and ones that use cooling to remove water from the air. When air passes by material that is cooled below the dew point of the air, water vapor condenses and collects on the cold material, leaving the air and making the air drier. Air conditioners and home dehumidifiers work by this same principle.

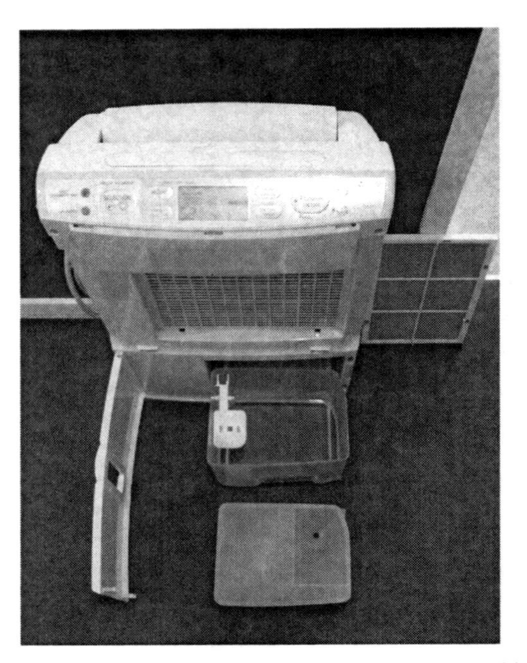

A home dehumidifier. Moist air passes through a filter and the chilled coils in the top of the machine. Water collects on the coils and drips into the water collection pan in the bottom of the machine.

Concept Reinforcement

1. What percentage of water vapor must be present to saturation air?
2. What is relative humidity?
3. What effects can humidity have on a system?
4. What is a desiccant?

Section 3.6 – Mass Balance Problems and Partial Saturation

Section Objective

- Solve material balance problems involving partial saturation

Solving Material Balance Problems Involving Partial Saturation

Solving material balance problems for partial saturation systems involves the same steps as solving a mass balance for any other system.

First of all, you have only water to deal with when solving a mass balance problem involving partial saturation. There are specific steps in performing a mass balance equation. These are:

1. Draw and label a process flow chart.

2. Select a basis for the calculation. This means that what units are you going to use in the calculation? How are the flow rates calculated? How is the mass presented?

3. Write a material balance equation.

Solve the equation for the unknown quantities.

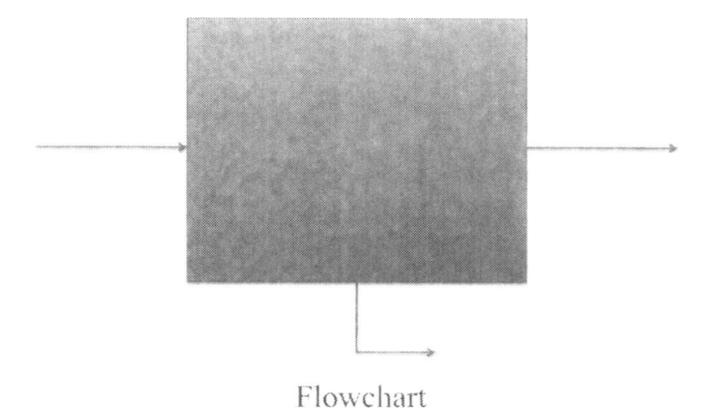

Flowchart

Application

Most examples of these types of calculations can be very complicated. For the sake of reinforcing mass balancing equations, we will simplify some of the calculations.

In our example we are looking at an air conditioner running in the state of Florida during the humid month of July. A typical air conditioner will create a relative humidity in the space at a range of 40-60% relative humidity. For this example, we will use the mean of 50%. If the relative humidity of the outside air is 80%, how much water vapor must be removed from the air?

An air conditioner works on the principle that hot humid air is pulled into the system where it passes over cool coils. These coils lower the temperature of the air, and cause the water vapor (humidity) to condense.

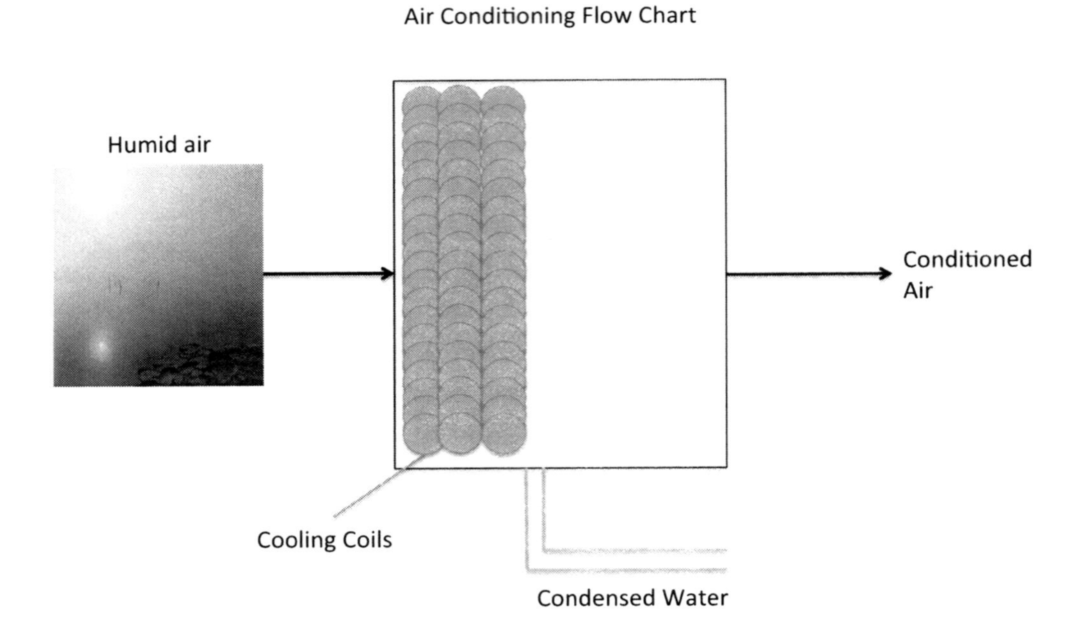

Based on the amount of water vapor in humid air we get the following values:

At 32 degrees C, there is 0.40 oz of water/lb of air at 80% humidity

At 32 degrees C, there is 0.25 oz of water/lb of air at 50% humidity.

If we are moving 1000 lbs of air through the air conditioner, how much water will end up in the condenser?

First, we must find out how much water is in the initial air and in the processed air.

Initial air: 1000 lbs of air · 0.40 oz of water/lb of air = 400 oz of water

Processed air: 1000 lbs of air · 0.25 oz of water /lb of air = 250 oz of water

We know that we are starting with 400 oz of water in the initial air After the air passes through the air conditioner we have 250 oz of water in the processed air

Water In = Water Out

Water In = (Condensed Water + Water in Processed Air)

400 oz = Condensed Water + 250 oz

Condensed Water = 150 oz

150 oz of water will end up in the condenser. If you ever look carefully at the outside part of a window air conditioner running in the summer heat, you will notice a lot of water is leaking out of the unit. This is the water that is being condensed out of the humid summer air. This explains why air conditioners are so nice in the summer – they both cool and dry the hot humid summer air.

Concept Reinforcement

1. What are the three steps to solving a mass balance problem?
2. What range of humidity is created indoors by an air conditioner?
3. In the example above, how much water would be condensed if 3,342 lbs of air moved through the air conditioner?

Section 3.7 – Gibbs Phase Rule

Section Objective

- State and explain the Gibbs Phase Rule

The Gibbs Phase Rule

The **Gibbs Phase Rule** is a rule that describes the possible number of degrees of freedom for a system in equilibrium. The rule can be expressed as:

$F = C - P + 2$ where:

F = **Degrees of Freedom**, which are the number of variables that need to be specified to determine the state of the system. These variables are usually temperature, pressure or composition.

C = **Chemical Constituents**, which are the distinct compounds or elements involved in the system that are in equilibrium.

P = **Phases**, which are the number of phases (i.e. gas, liquid or solid) present in the system. These phases are not miscible in each other and can be mechanically separated.

An iceberg: ice floating in water

Example 1: If we are using this rule for a pure substance, C=1 therefore; $F = 3 - P$. If there is a single phase, then $P = 1$ and F=2. That would be two variable to be stated such as temperature and pressure. This system can exist in several different temperature and pressure ranges.

Example 2: If we look at water, which can exist as ice, liquid water and water vapor, we could have three phases for one constituent. That would give us the following equation:

$$F = 1 - 3 + 2$$

$$F = 0$$

With $F = 0$, this means that three phases of one constituent could only happen at a single temperature and pressure. This single pressure and temperature combination is known as the triple point for water. The **triple point** is the only pressure and temperature combination in which the solid, liquid, and gas phases of a substance can all coexist in equilibrium. Any variation in temperature or pressure and you will not have three phases in equilibrium.

Steam rising from a hot geyser pond

Concept Reinforcement

1. What does the Gibbs Phase Rule describe?
2. What happens when the degrees of freedom are 0?
3. What is special about the triple point of a substance?
4. A tank contains a mixture of liquid helium and helium gas. How many degrees of freedom does the system have?

Section 3.8 – Vapor-Liquid Equilibria in Binary Systems

Section Objective

- Explain vapor-liquid equilibria in binary systems

Vapor-Liquid Equilibrium

Vapor-liquid equilibrium is the movement of liquid and gas back and forth from the liquid to the vapor phase. What is happening is that the rates of condensation and evaporation are the same. Vapor-liquid equilibrium is most commonly used in chemical engineering in the process of distillation.

During distillation, liquid mixtures are heated so that they can be separated based on differences in the boiling points of the liquids in the mixture. The liquid with the lowest boiling point will evaporate at a lower temperature than liquids in the mixture with higher boiling points. As the mixture is heated, all of the liquids in the mixture will move back and forth from their liquid and vapor phases, but some of the liquid with the lowest boiling point stays in the vapor phase. This vapor is cooled in a condenser and collected in a flask to separate it from the other liquids in the mixture.

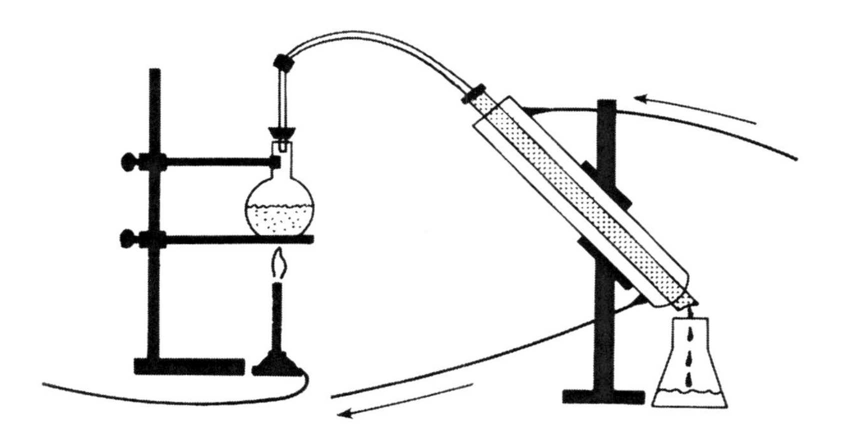

Basic distillation apparatus

Binary Systems

A **binary system** is a system that has two components. The concentrations of each component are compared in the liquid and gaseous phase. The concentrations are often expressed as mole fractions for each component in each of the phases.

In these binary systems:

$$P_{liq} = P_{gas}$$

$$T_{liq} = T_{gas}$$

$$G_{liq} = G_{gas}$$

Where P = pressure, T = temperature and G = Gibbs Free Energy. Gibbs Free Energy is a measure of the units of energy per amount of substance. There will be one Gibbs Free Energy factor for each constituent.

In binary systems as you alter the pressure and the temperature, the mole fraction of each constituent in the liquid phase varies based on the pure substance's boiling points. Now if you hold the pressure constant, you can compare mole fractions of the two components at various temperatures. This is called a **boiling point diagram**.

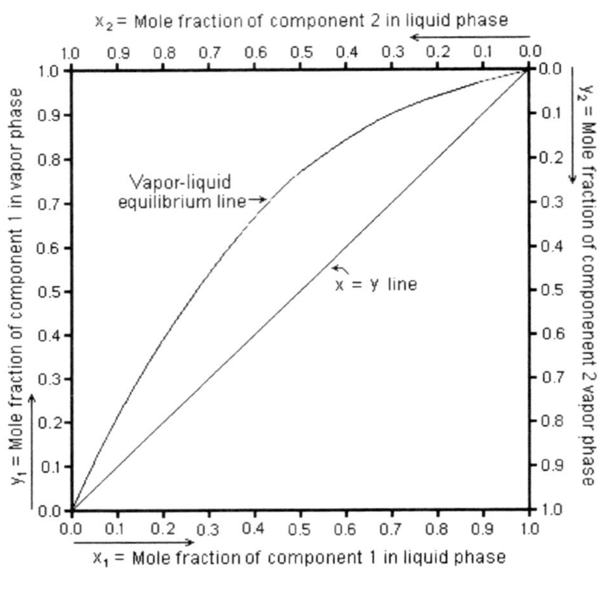

Boiling point diagram

Boiling Point Diagram

In a boiling point diagram you plot the two mole fractions against a changing temperature. When the points are plotted against the different temperatures, a graph with a bubble appears. The two end points are the boiling points of the two pure substances. The bottom curve is called **bubble point curve** and it represents mole fraction of the boiling liquid at various temperatures. The upper curve is called the **dew point curve** and it represents the vapor-liquid equilibrium at various temperatures.

Concept Reinforcement

1. What is the rate of condensation the same as in vapor-liquid equilibria?
2. What is a binary system?
3. In a boiling point diagram what does the upper curve represent? What is it called?
4. In a boiling point diagram what does the lower curve represent? What is it called?

Section 3.9 – Liquid and Gases in Equilibrium with Solids

Section Objective

- Describe liquids and gases in equilibrium with solids

Solid-Gas Equilibrium

Solid-gas equilibrium can exist just as liquid-gases equilibrium does. As a solid converts to a gas, the process is called **sublimation**, while the reverse process of gas to solid is called **deposition**.

Like the liquid-gas equilibrium, the solid-gas equilibrium is related to temperature. As the temperature increases, the amount of gas generated increases. As temperature is decreased, the substance moves toward the solid.

An example of this type of equilibrium is dry ice, or solid carbon dioxide. If this substance is placed in a sealed Erlenmeyer flask at the right temperature, you will see a vapor fill the flask but not lose all the solid carbon dioxide.

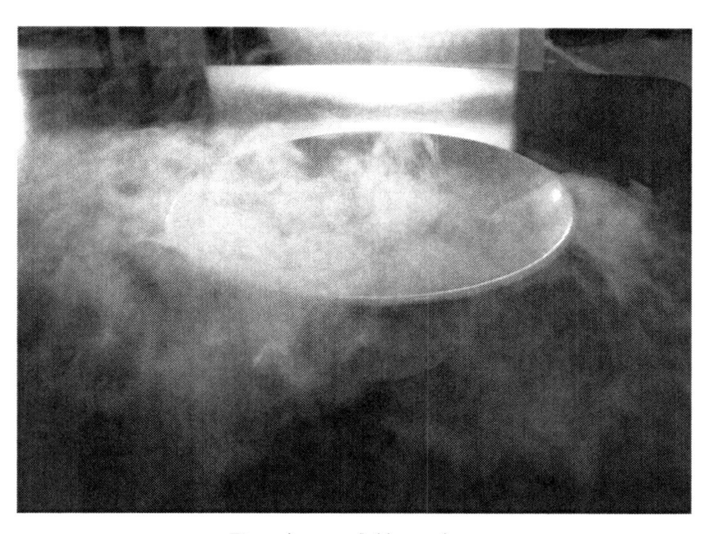

Dry ice sublimation

Solid-Liquid Equilibrium

Solid-liquid equilibrium can exist just as liquid-gases equilibrium does. As a solid converts to gas, the process is called **melting**. As the process reverses, the process is called **freezing**.

Like the liquid-gas equilibrium, the solid-liquid equilibrium is related to temperature. The amount of liquid formed from solid increases as the temperature increases. As temperature is decreased, the substance moves toward the solid.

An example of this type of equilibrium is ice, or solid water. If this substance is placed in a sealed Erlenmeyer flask at the right temperature, you will see a liquid start to form on the solid. As long as the pressure and temperature are maintained, the amount of water and solid together will not change.

Ice melting on Lake Superior

Concept Reinforcement

1. What is the process of a solid converting to a gas called?
2. What is the process of a solid converting to a liquid called?
3. What is the process of a gas converting to a solid called?
4. What is the process of a liquid converting to a solid called?

Section 3.10 – Energy Balances and Types of Energy

Section Objective

- Discuss energy balances and types of energy

Energy Balance

Energy balance is a method of quantifying energy used or produced by a system. The **Law of Conservation of Energy** states that energy cannot be created or destroyed. Because energy cannot be created nor destroyed, it is possible to account for all the energy, especially if it is in an isolated system. In simple terms, the amount of energy that goes in is the same as the amount of energy that comes out.

Types of Energy

While energy cannot be created or destroyed, it can change from one form of energy to another. There are many different types of energy. Some of the more common types are:

Electromagnetic energy, which is also called electromagnetic radiation, is the energy that is transmitted by waves through space or some other medium. Types of electromagnetic energy are radio, x-ray, microwaves, gamma rays and visible light.

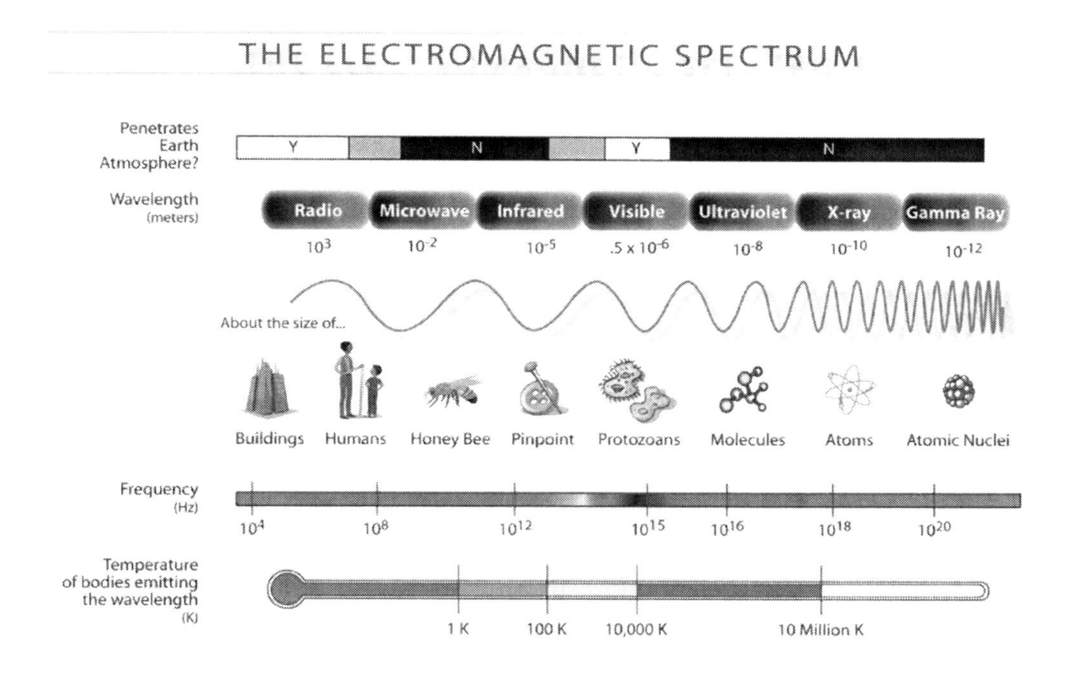

The electromagnetic spectrum

Image courtesy of NASA

Kinetic energy is the energy that an object obtains due to its motion. Think of a car rolling down a hill. The energy from that movement is kinetic energy. Molecules may also have kinetic energy as well, which we usually think of as temperature.

Potential energy is the energy that is stored is a system. Think of a ball sitting at the top of a hill. The energy that is contained in the ball before it rolls down the hill is considered potential.

Thermal energy, also known as **heat**, is the energy that is created during a rise in temperature. Thermal energy is a measure of the kinetic energy possessed by the molecules that make up a substance. Think of a pot on the stove. The heat from the stove raises the temperature of the pot. Most energy that is "wasted" in a chemical process is changed to thermal energy. This explains why machinery tends to heat up as it is used – the heat is energy that is leaving the system as waste.

Chemical energy is the energy released or absorbed as a result of a chemical reaction. Molecules store energy in their chemical bonds that can be released when the bonds are rearranged. Some chemical reactions release chemical energy as heat or light, while others absorb thermal energy to store it as chemical energy and appear to become "colder" as a result of their heat absorption.

Nuclear energy is the energy released from nuclear fission or fusion reactions. This energy is released from the forces that hold subatomic particles together in an atom.

Fire converts the chemical energy stored in the chemical bonds of wood into electromagnetic energy (light) and thermal energy (heat)

Concept Reinforcement

1. What is the basic premise of energy balance?
2. What is kinetic energy?
3. What is another term for potential energy?
4. What is thermal energy?
5. Why is thermal energy important in chemical engineering?

Section 3.11 – Conservation of Energy

Section Objective

- Explain the concept of Conservation of Energy

Conservation of Energy

The **Law of Conservation of Energy** is a law that states that energy cannot be created or destroyed. In an isolated system the amount of energy in the system is always constant. It can change from one form of energy to another, but it cannot be destroyed or created. As with other conservation laws such as the Conservation of Mass, the energy that goes into a system must come out.

Energy In = Energy Out

The Law of Conservation of Energy has many direct implications in engineering. Since the time of the Industrial Revolution, scientists and engineers have searched for a way to create a perpetual motion device. Some type of device that would work indefinitely once is gets started. Many of you have seen the little toy is a series of balls suspended by a string. You swing the first ball out, and it hits the line of balls, sending the last ball up in an arc. The last ball swings back, hitting the same line of balls sending the first ball out in an arc. This seems like it would go on forever, but due to forces such as friction and gravity, the device slows down. Each ball that swings in an arc swing a little less than the previous time until it stops. In this example the kinetic energy has been transformed into thermal energy by friction between the balls.

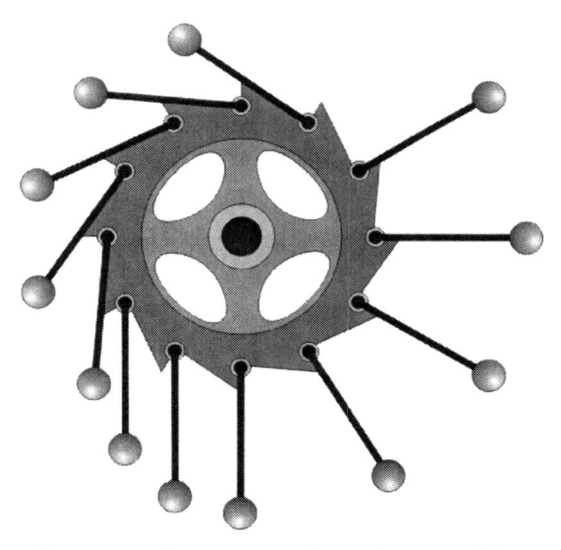

Cartoon of a perpetual motion machine

Concept Reinforcement

1. What does the Law of Conservation of Energy state?
2. If energy cannot be created or destroyed, what can happen to it?
3. Why is it impossible to build a perpetual motion machine?

Section 3.12 – Energy Balances for Closed Systems

Section Objective

- Describe energy balances for closed, unsteady- and closed, steady state systems

Energy Balance

According to the Law of Conservation of Energy, energy can neither be created nor destroyed. It can be changed to other forms, and cross boundaries of a system, changing the total energy of the system. In its simplest terms:

Energy In = Energy Out

In a system to balance the energy, you have the following equation:

Total Energy + Convection Energy = Heat into the System + Work Done on the System

Total Energy is of course the total energy of the system. **Convection Energy** is the net flow of energy out of the system.

In order to balance the energy of a system, we need to know all the types of energy that are in a system. Is there movement in the system? If so there will be changes in kinetic or potential energy? With movement there is work. Does this work generate heat that might leave the system? All these energies must be accounted for in the energy balance.

If there is there are no mechanical parts in the system, and the system cannot expand, there can be no work performed in the system.

Closed Systems

A closed system is a system that is isolated from the surrounding environment. There is no mass coming in or out of the system, but energy can move across the boundaries. So to balance a closed system you need to calculate the change in energy form the initial conditions of the system to the final conditions in the system. These changes occur to energy being brought into the system such as heat and work. Energy can also leave the system as convection.

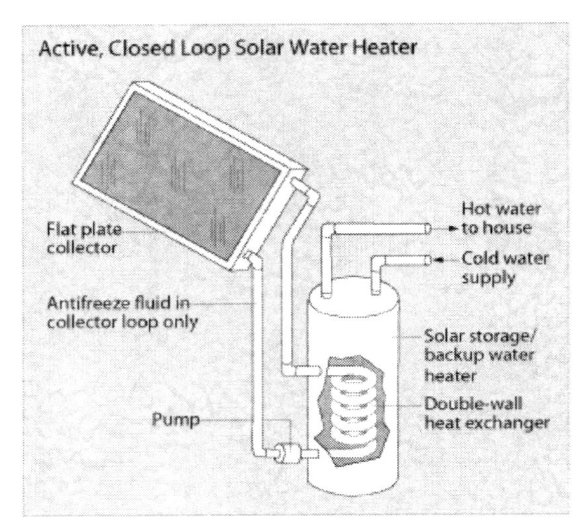

Image courtesy of US Department of Energy

Steady-state Systems

A steady-state system is a system where the variables do not change with time. Being a closed system, the amount of matter remains the same, but the energy as stated can cross boundaries. So to describe energy balance in a steady state system it is similar. The energy or energy change, by definition is constant throughout the system. Therefore as long as you have identified all types of energy in the system, the type and amount of each type of energy will remain constant and balancing should be easy.

Unsteady-state Systems

An unsteady-state system is a system where the variables are constantly changing. Since the system is closed, the mass remains the same as it does in a steady-state system, but the amount of energy is the system can fluctuate depending on the changing conditions in the system. As an engineer you need to keep track of these changing energy ratios over time. Since it is a closed unsteady-state system, the system will tend to move towards thermodynamic equilibrium.

Concept Reinforcement

1. In energy balance, what is the amount of energy flowing into the system equal to?
2. What is the flow of energy out of a system called?
3. What defines a closed system?
4. Describe the variables of a steady-state system.

Section 3.13 – Energy Balances for Open Systems

Section Objective

- Describe energy balances for open, unsteady- and closed, steady state systems

Energy Balance

According to the Law of Conservation of Energy, energy can neither be created nor destroyed. It can be changed to other forms, and cross boundaries of a system, changing the total energy of the system. In its simplest terms:

Energy In = Energy Out

The following equation shows the balance of energy in a system:

Total Energy + Convection Energy = Heat into the System + Work Done on the System

Total Energy is of course the total energy of the system. Convection Energy is the net flow of energy out of the system.

In order to balance the energy of a system, we need to know all the types of energy that are in a system. Is there movement in the system? If so there will be changes in kinetic or potential energy? With movement there is work. Does this work generate heat that might leave the system? All these energies must be accounted for in the energy balance.

If there is there are no mechanical parts in the system, and the system cannot expand, there can be no work performed in the system.

The Rancho Seco nuclear power plant in Sacramento County, California, USA

Open Systems

An open system is a system where variables like mass, energy and momentum can move through boundaries. As indicated by its name, an open system is open to the environment. Energy and mass can be added to the system at any time. In closed systems where we were only concerned about initial and final conditions, with open systems we have to monitor all systems at all times.

Steady-state Systems

As we know, a steady-state system is a system where the variables do not change with time. In an open system, energy is entering and leaving the system at a constant rate. Think of a power plant. Fuel enters the burner at a constant rate, is burned which generates steam. The steam is at a constant temperature and is used to turn a turbine that creates electricity. This system is a steady state since the variables do not change. You have fuel (energy) coming into the system, heat (burning), work (turning the turbine) which generated electricity which is energy leaving the system.

As we have shown, there are four identified energies. Others exist such as friction on the generator, kinetic energy for the fuel being injected into the burner as well as the steam moving through the pipes. There is also energy that is lost up the smoke stack. So as you see there are many different areas to maintain an energy balance, but they are constant and therefore easy to track in a steady-state system.

The Mount Storm coal power plant in Grant County, West Virginia, USA

Unsteady-state Systems

As we know, unsteady-state system is a system where the variables are constantly changing. As an open system, there is energy and mass entering the system at various rates.

An example of an open unsteady-state system is a commercial beer production. In beer production, the ingredients are added to a large tank, and boiled (heat added). The material is transferred through a chiller where the heat is removed (energy out) on its way to the fermenter. Here the beer cools more (energy loss) but the fermentation begins. Here the compounds in the beer are broken down into carbon dioxide and alcohol. The breaking of bonds is also a release of energy. As you can see, this process is not steady. Each step causes changes to the total energy of the system at varying rates. Therefore balancing the energy equation for an open unsteady-state system is difficult.

Brewing vats

Concept Reinforcement

1. In an energy open system, what variables can move through boundaries?
2. What can be said about energy entering and leaving a steady-state system?
3. What is the most difficult system to perform an energy balance on?

Section 3.14 – Enthalpy

Section Objective

- Explain enthalpy changes

What is Enthalpy?

Enthalpy is the heat content of a chemical system. It was discovered that almost all chemical reactions either absorb or release heat, therefore all substances contain heat.

Welding rails using the thermite reaction – a very exothermic chemical reaction

What is Enthalpy Change?

Enthalpy change is the release or absorption of heat in a chemical reaction. This difference in heat is calculated as:

Change in Enthalpy = Enthalpy (products) – Enthalpy (reactants)

This equation can also be written as:

$$\Delta H = H_{products} - H_{reactants}$$

where

ΔH is the change in enthalpy in kilojoules (kJ)

$H_{products}$ is the enthalpy of the products (in kJ)

$H_{reactants}$ is the enthalpy of the reactants (in kJ)

If the result of this equation is that the change in enthalpy is negative, then heat is released and the reaction is said to be **exothermic**. A negative change in enthalpy means that there is less energy in the system after the chemical reaction, so the energy must have been released as heat.

If the result of this equation is that the change in enthalpy is positive, then heat is absorbed and the reaction is said to be **endothermic**. A positive change in enthalpy means that there is more energy in the system after the chemical reaction, so the energy must have been absorbed by the products from their surrounding environment.

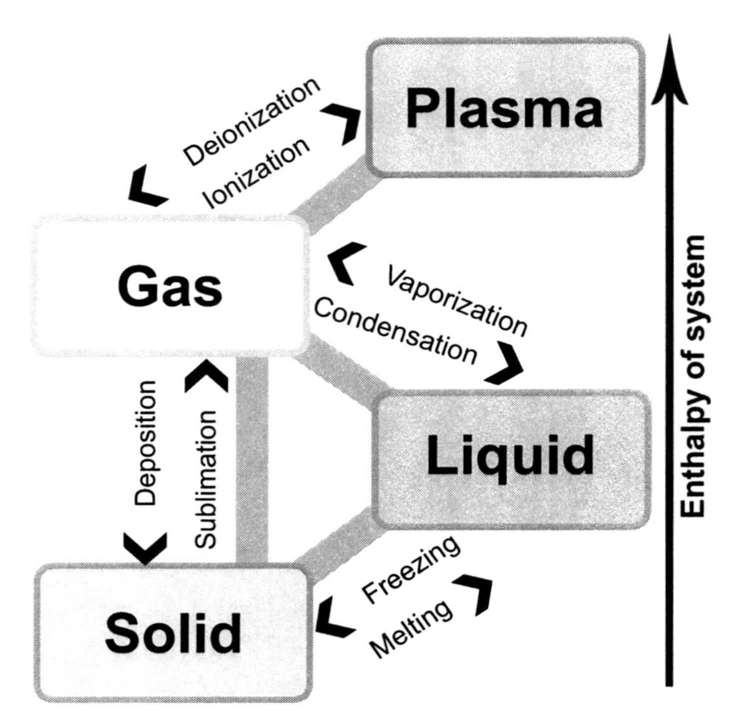

In an exothermic reaction, the energy of the reactants is greater than the energy of the products. For an endothermic reaction, the reverse is true. The energy of the products is greater than the energy of the reactants.

Example: A chemical reaction takes place between compound A and compound B to form compound C. If the enthalpy of the reactants is -500 kJ and the enthalpy of the product is -600 kJ, find the change in enthalpy and decide if the reaction is endothermic or exothermic.

$\Delta H = H_{products} - H_{reactants}$

ΔH = the change in enthalpy in kilojoules (kJ)

$H_{products}$ = -600 kJ

$H_{reactants}$ = -500 kJ

ΔH = -600 kJ – (-500 kJ)

ΔH = -600 kJ + 500 kJ = -100 kJ

The enthalpy change is -100 kJ. Because this number is negative, it tells us that the reactants have more energy than the products, so the reaction must be exothermic – it releases the extra energy into the environment.

There are several types of enthalpy changes:

1. **Heat of Reaction** is the enthalpy change during a chemical reaction

2. **Heat of Formation** is the enthalpy change that occurs when a compound is formed.

3. **Heat of Fusion** is the enthalpy change during a phase transition between a solid and a liquid. For example, the heat of fusion is the heat absorbed when ice melts. This explains why ice is so effective in cooling drinks.

4. **Heat of Vaporization** is the enthalpy change during a phase transition between a liquid and a gas. For example, when you sweat, the sweat absorbs energy as it evaporates from your skin. This results in you feeling cooler.

Enthalpy and Chemical Engineering

As we have seen, enthalpies can be either positive or negative, resulting in variations in the amount of heat produced. How is this applicable to a chemical engineer? When looking at a process knowing how much heat is produced in a reaction is very important. If too much heat is being generated, the reaction may get violent, destroying equipment and possible injuring people. If a reaction absorbs too much heat, the reaction may be expensive to run if more and more energy must be added to the system to keep the reaction going.

Concept Reinforcement

1. What is enthalpy?

2. What do you call a reaction that generates heat?

3. What do you call a reaction that has a positive enthalpy?

4. What type of enthalpy change reaction occurs when ice melts?

5. A chemical reaction takes place between compound A and compound B to form compound C. If the enthalpy of the reactants is -355 kJ and the enthalpy of the product is -325 kJ, find the change in enthalpy and decide if the reaction is endothermic or exothermic.

Section 3.15 – Applications of Chemical Engineering

Section Objective

- Describe an application of Chemical Engineering

Application of Chemical Engineering: Wind Turbine

Chemical engineers are able to work in a wide variety of jobs throughout many industries. Although the focus of most chemical engineers work is on the chemical plant or oil refinery, many chemical engineers choose to apply their skills and knowledge to fields that do not deal strictly with the production of chemicals. A good application of the theories studied by chemical engineers involving energy balances is a wind turbine, also known as a wind mill.

As the cost of energy and society's demand for energy increases, many people are looking to wind power to solve our energy problems. Wind power is a clean and environmentally friendly, or "green," energy source. Wind is also free and does not use up any raw materials. The cost of wind energy is in the construction, maintenance, and management of the wind turbines that collect energy from the wind.

At the USDA-ARS Conservation and Production Research Laboratory in Bushland, Texas, wind turbines generate power for submersible electric water pumps that are far more efficient than traditional windmills (background).

Image courtesy of the USDA

In a large system like a wind farm you have several separate systems to identify. To easily balance the energy in this system, you need to separate it into separate units and calculate an energy balance for each.

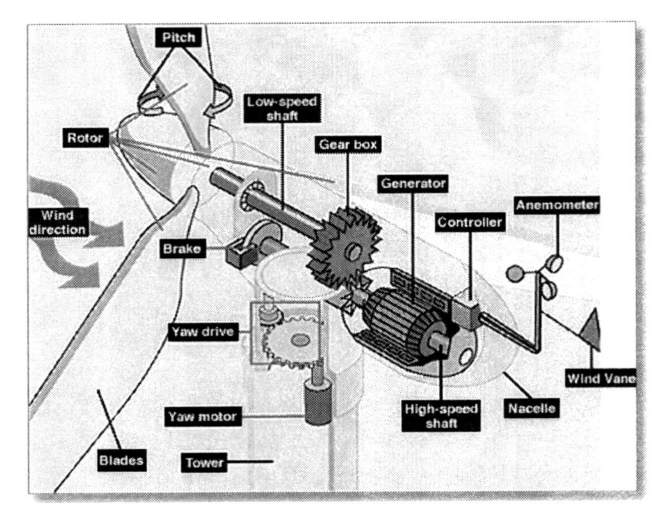

The components of a wind turbine

First of all you have the wind turbine. This captures the kinetic energy of the wind to turn the turbine. This is a closed system. If the wind is steady, the system is in steady-state. If the wind is irregular the system is in unsteady-state.

The next step is the generator. The energy from the wind turns the turbine which in turn, turns the generator creating an electrical current. In this system, the energy from the wind is converted into electricity. Another type of energy that you have to account for is heat. If you have ever felt a generator on a car or a toy, you can feel the heat that it gives off. This is part of the energy balance. Since there is no mass entering or leaving the system, the system is closed. Since the wind is not consistent, the system is unsteady-state.

Next we have a battery. The battery is used to store excess energy form the turbine. If the turbine is spinning, energy is going into the battery, but not coming out. When the turbine is still, the battery sends out electricity. Therefore the system is closed, but unsteady-state.

The next step in the systems is the electrical grid. In this system the electrical energy created from the wind is distributed through the power lines. Since all the electricity is used, the system is closed. In addition, because the combination of the battery and the wind makes the electricity consistent, the system is in steady-state.

So as an engineer, you take into account all the various energies impacting this system and calculate them. By ensuring that everything is accounted for, you can increase the efficiency of the system.

Concept Reinforcement

1. What is the first thing that is done to calculate the energy balance on a large complex system?
2. Generators create electricity, but what other type of energy do they also produce?
3. What makes a wind turbine ultimately a steady-state system?

Appendix

Chemical Engineering Answer Key – Unit 1

Section 1.1

1. Scientists develop a body of knowledge about a topic by doing experiments. Scientists do experiments using the scientific method, which is a disciplined, systematic approach to developing and answering questions, to add to the body of knowledge.

2. Engineers use scientific principles to develop the tools and products used by people just like you. Engineers design buildings, cars, eyeglasses, furniture, electronics, cleaning products, medications, medical equipment and many other things.

3. a. Science
   b. Engineering
   c. Engineering
   d. Science
   e. Science
   f. Engineering

Section 1.2

1. A unit is a set amount of a measurement.

2. 3.160 km

3. 0.050 L

4. 45.42 L

5. 3.048 m

Section 1.3

1. An atom is the building block of all matter.

2. The Greeks

3. Ernest Rutherford

4. Quarks

5. a. Positive
   b. Neutral (no charge)
   c. Negative

Section 1.4

1. An element is a pure chemical substance composed of atoms with the same number of protons.

2. Alkali metals

3. Noble gases

4. Atomic radius

5. The periodic table organizes the elements based on their atomic numbers, physical and chemical properties, and trends like electronegativity, atomic radius, ionization energy, and electron affinity.

Section 1.5

1. Molecules are the smallest amount of a substance that still has the properties of that substance. A molecule may be a pure substance or may be a compound. Compounds are special types of molecules that are made of 2 or more different types of elements.

2. Compounds are molecules made from 2 or more elements that are chemically bonded. Mixtures are made from 2 or more substances that are physically or mechanically combined, but not chemically bonded.

3. Hydrogen is a molecule.

4. Carbon dioxide is a compound.

5. You have a mixture.

Section 1.6

1. A chemical bond is the attractive interaction between atoms.

2. An ion is an atom or molecule that has either gained or lost electrons.

3. The electrons are donated from one atom to another to form ions in an ionic bond.

4. a. Fluorine
   b. Lithium
   c. Fluorine

Section 1.7

1. Covalent bonds involve the sharing of electrons between 2 atoms. Ionic bonds involve the donation of electrons from one atom to another.
2. Covalent bonds form between elements with similar electronegativities.
3. A triple bond is stronger than a single bond.
4. A covalent bond is usually stronger than an ionic bond.

Section 1.8

1. Carbon 12
2. 6.02×10^{23}
3. 107.8682
4. 6.941 g/mol

Section 1.9

1. Both objects would have the same density.
2. 40 g/mL
3. The piece of metal has a specific gravity of 20. The piece of metal would sink when placed in water.
4. 60 m^3/s

Section 1.10

1. 5,000 ppb
2. 0.5 ppm
3. 0.1 kg/L
4. 0.7 M

Section 1.11

1. Qualitative analysis detects the presence of a substance while quantitative analysis determines how much of a substance is present in a sample.

2. Gravimetric

3. Chromatography (gas or liquid) or electrophoresis

4. Electrophoresis

Section 1.12

1. A basis is a starting point, a foundation, or the basics that you need to start or perform a project.

2. It is important to have a basis before beginning a production process so that everything is planned out, including the units of measurement used in the process.

3. The units of measurement to be used ; the schematics of the machinery; the materials needed; the outputs from the process; safety considerations; pollution control strategies; laws concerning pollution, noise, worker safety, and zoning of property; redundant and backup systems; how much product will be produced per hour, day, etc.; the estimated cost of the product

4. Mathematical errors can occur when units have to be converted from one system to another throughout a production process.

Section 1.13

1. The Fahrenheit scale

2. 373.15 K

3. 323.15 K

4. 122 °F

Section 1.14

1. Pounds per square inch (lb/in^2 or psi) and Pascals (Pa)

2. The pressure of the air decreases as altitude increases.

3. 110 Pascal

4. 200 psi

Section 1.15

1. Petroleum is crude oil.

2. Petroleum is separated into its components by fractional distillation during the refining process.

3. Chromatography and spectroscopy are used to measure the purity of the final product.

4. Gasoline, kerosene, diesel fuel, heating oil, motor oil, tar, asphalt, natural gas, and petrochemicals.

Chemical Engineering Answer Key – Unit 2

Section 2.1

1. The Law of Conservation of Matter.

2. Chemicals are lost as pollutants.

3. The reactants were not combined in the correct proportion.

4. 20 kg

Section 2.2

1. The human body is an open system because materials move into the system (food, water, and air) and out of the system (wastes, water, and air).

2. Part of the pharmaceutical industry, refrigeration loops, water systems in nuclear reactors, solar water heaters, and automobile coolant systems are some of many different examples.

3. It is rare to have a closed system because materials are being added and products being removed to make money from the industrial plant.

Section 2.3

1. The manufacturing of steel cable
2. The making of wine
3. The variables in a steady-state system stay the same while the variables in an unsteady-state system change over time.
4. It is important to keep all of the variables the same to produce a consistent product and minimize waste.

Section 2.4

1. Screening to physically separate a mixture, combustion to produce sulfuric acid, steam production to generate power, chemical blending, chemical purification, chemical drying, and many other possible answers.
2. Multiple component systems are more common than single component systems because most chemical processes are very complicated and require multiple steps.
3. Each component has an output. Some of these outputs can be used to make different products.
4. If one step has an error, it can lead to errors in other steps as well and ultimately a bad product.

Section 2.5

1. Reactants
2. By-products
3. a. $4Al + 3O_2 \rightarrow 2Al_2O_3$
   b. $2NaOH + H_2SO_4 \rightarrow Na_2SO_4 + 2H_2O$
   c. $4NH_3 + 3O_2 \rightarrow 2N_2 + 6H_2O$

Section 2.6

1. Baking bread, among other examples.

2. A car engine, among other examples.

3. Fermentation, among other examples.

4. 0 kg

Section 2.7

1. 1. Draw and label a process flow chart.
   2. Select a basis for the calculation. This means that what units are you going to use in the calculation? How are the flow rates calculated? How is the mass presented?
   3. Write a material balance equation.
   4. Solve the equation for the unknown quantities.

2. 206.19 kg of H_2SO_4 and 1793.81 kg of water (rounded to the nearest 0.01 kg).

3. a. 83.33 kg of NaClO (rounded to the nearest 0.01 kg)
   b. 416.67 kg of water (rounded to the nearest 0.01 kg)
   c. It will take 0.83 hours to make 500 kg (rounded to the nearest 0.01 kg).

Section 2.8

1. $Fe + 2HCl \rightarrow FeCl_2 + H_2$

2. $2Al + 6HCl \rightarrow 2AlCl_3 + 3H_2$

3. $3Ag + 4HNO_3 \rightarrow 3AgNO_3 + 2H_2O + NO$

4. Stoichiometry allows chemical engineers to use the correct ratio of reactants in a reaction.

Section 2.9

1. $2N_2 + 5O_2 + 2H_2O \rightarrow 4HNO_3$

2. 25 moles of nitrogen gas and 62.5 moles of oxygen gas.

3. $4Al + 3O_2 \rightarrow 2Al_2O_3$

4. 200 moles of aluminum and 150 moles of oxygen gas.

5. $2C_2H_6 + 7O_2 \rightarrow 4CO_2 + 6H_2O$

6. 100 moles of ethane and 350 moles of oxygen gas

Section 2.10

1. 1. Draw a flow chart and label all streams.
   2. Select a basis for the calculation. This means that what units are you going to use in the calculation? How are the flow rates calculated? How is the mass presented?
   3. Separate each system by drawing a box around it. This will allow you to calculate each system separately.
   4. Write a material balance equation for each separate system.
   5. Solve the equation for the unknown quantities.

2. a. 360 kg
   b. 540 kg
   c. 40 kg

3. 75 kg of Chemical C and 355 kg of Final Product

Section 2.11

1. Physical processes

2. Washing, distillation, drying, evaporation, crystallization, absorption, extraction, adsorption, or any other physical process

3. Ore comes into the plant and is washed to remove any sand that might be in the ore. The slurry of sand and water is allowed to separate, and the water is recycled into the process to be used again to wash ore. The recycling of water reduces the amount of fresh water used by the plant and reduces the amount of dirty waste water that leaves the plant.

4. Recycling allows more product to be made with less materials, keeps costs low, and reduces waste and pollution.

Section 2.12

1. Help increase the efficiency of the process
   Help to speed up the process
   To improve separation
   To minimize cost of additional equipment and supplies

2. A splitter removes unused reactants from products so that the reactants can be recycled back into the reactor.

3. There is always some matter lost as waste due to the nature of atoms and molecules being so small that they sometimes fail to mix with other particles in a reaction.

Section 2.13

1. An engineering process that skips one or more subsequent processes.

2. Bypasses can be used to allow the relief of pressure in a process where pressure can build up too high. Bypasses can be used to introduce new or additional reactants later in a reaction. They can also be used to obtain precise control of the final product by adjusting how much material goes through each process. In some cases, the same equipment can be used to make several different products. Bypass can be used in combination with recycle to pull finished product from a recycle loop.

3. Reactants that are obtained from nature, like minerals mined from the ground or agricultural products like grain tend to vary in their quality. Bypass can be used to alter the amount of processing needed to use natural materials.

4. Bypass can be used in combination with recycle to pull finished product from a recycle loop.

Section 2.14

1. 1. Bleeding off from the process stream.
   2. To make a tank or vessel inert.
   3. To keep a recycle system in balance.

2. Liquids are purged through a drain.

3. If too much material is recycled, pressure can build to dangerous levels.

4. Petroleum vapors left in an empty tank can explode.

Section 2.15

1. Batch

2. Coking and cracking allow large hydrocarbons to be broken down into high demand hydrocarbons like gasoline.

3. Plastics are cheap, flexible, durable, and can be made into many different forms.

4. Plastics are made from petrochemicals that are made from oil by refineries.

5. Plastics can be shaped by injection molding or extrusion.

Chemical Engineering Answer Key – Unit 3

Section 3.1

1. 1. The space that a gas molecule occupies is significantly smaller than the space between the gas molecules.
   2. The intermolecular forces between the gas molecules are negligible.

2. High pressure and low temperature.

3. 15.52 mol (rounded to the nearest 0.01 mol)

4. 147.22 L

Section 3.2

1. Real gases do not follow the same rules as ideal gases. Real gases exist at all pressures and temperatures.

2. Boyles Law and Charles and Gay-Lussac's Law make up the ideal gas law.

3. The van der Waals equation is applicable to real gases and includes variables to account for the fact that each gas act has a different way of acting at high pressures and low temperatures.

4. 1.79 mol of oxygen

Section 3.3

1. Vapor pressure is the equilibrium between the gaseous and solid or liquid phase of a substance.

2. Sublimation

3. Evaporation

4. A = Triple point
   B = Liquid/solid equilibrium curve
   C = Liquid/gas equilibrium curve or the vapor pressure curve
   D = Solid/gas equilibrium curve

Section 3.4

1. A sponge full of water is one of many examples of saturation.

2. The dew point

3. Water evaporating from a tea kettle on a stove is one of many examples of vaporization.

Section 3.5

1. 100%

2. Relative humidity is the ratio of the partial pressure of water vapor to the saturated vapor pressure of water vapor.

3. Humidity can affect a chemical reaction, damage equipment, and can affect the safety of people working in a chemical plant.

4. A chemical that absorbs water.

Section 3.6

1. 1. Draw and label a process flow chart.
   2. Select a basis for the calculation. This means that what units are you going to use in the calculation? How are the flow rates calculated? How is the mass presented?
   3. Write a material balance equation.

2. 40-60% relative humidity

3. 501.3 oz of water.

Section 3.7

1. Gibbs Phase Rule describes the possible number of degrees of freedom for a system in equilibrium.

2. The substance is at its triple point, and neither temperature nor pressure can be changed without taking the system out of equilibrium.

3. The triple point is the only temperature and pressure combination at which a substance can exist as a solid, liquid, and gas at equilibrium.

4. 1

Section 3.8

1. The rate of evaporation.

2. A system that has 2 components.

3. The upper curve represents the vapor-liquid equilibrium at various temperatures. It is called the dew point curve.

4. The lower curve represents the mole fraction of boiling liquid at various temperatures. It is called the bubble point curve.

Section 3.9

1. Sublimation

2. Melting

3. Deposition

4. Freezing

Section 3.10

1. The amount of energy that goes in is the same as the amount of energy that comes out.

2. The energy that an object obtains due to its motion.

3. Stored energy

4. Heat

5. Most of the energy that is lost as waste is lost as heat. Also, equipment cannot be allowed to overheat.

Section 3.11

1. Energy cannot be created or destroyed.
2. Energy can change forms.
3. The forces of friction and gravity slow down the motion of the machine.

Section 3.12

1. The amount of energy flowing out of the system.
2. Convection energy
3. A system that is isolated from its surrounding environment with no mass moving into or out of the system.
4. The variables of a steady-state system stay constant all of the time.

Section 3.13

1. Mass, energy, and momentum
2. Energy is entering and leaving the system at a constant rate.
3. Unsteady-state systems

Section 3.14

1. Enthalpy is the heat content of a chemical system.
2. An exothermic reaction
3. Endothermic
4. Heat of fusion
5. 30 kJ, endothermic

Section 3.15

1. To easily balance the energy in this system, you need to separate it into separate units and calculate an energy balance for each.

2. Heat, or thermal energy

3. The combination of the battery and the wind makes the electricity consistent, so the system is in steady-state.